NEW DEVELOPMENTS IN MEDICAL RESEARCH

DYSBIOSIS

A STUDY OF UNDERLYING CAUSES

NEW DEVELOPMENTS
IN MEDICAL RESEARCH

Additional books and e-books in this series can be found on Nova's
website under the Series tab.

NEW DEVELOPMENTS IN MEDICAL RESEARCH

DYSBIOSIS

A STUDY OF UNDERLYING CAUSES

RICHARD I. COWELL
EDITOR

nova
Medicine & Health
New York

We have partnered with Copyright Clearance Center to make it easy for you to obtain permissions to reuse content from this publication. Simply navigate to this publication's page on Nova's website and locate the "Get Permission" button below the title description. This button is linked directly to the title's permission page on copyright.com. Alternatively, you can visit copyright.com and search by title, ISBN, or ISSN.

For further questions about using the service on copyright.com, please contact:
Copyright Clearance Center
Phone: +1-(978) 750-8400   Fax: +1-(978) 750-4470   E-mail: info@copyright.com.

NOTICE TO THE READER

The Publisher has taken reasonable care in the preparation of this book, but makes no expressed or implied warranty of any kind and assumes no responsibility for any errors or omissions. No liability is assumed for incidental or consequential damages in connection with or arising out of information contained in this book. The Publisher shall not be liable for any special, consequential, or exemplary damages resulting, in whole or in part, from the readers' use of, or reliance upon, this material. Any parts of this book based on government reports are so indicated and copyright is claimed for those parts to the extent applicable to compilations of such works.

Independent verification should be sought for any data, advice or recommendations contained in this book. In addition, no responsibility is assumed by the Publisher for any injury and/or damage to persons or property arising from any methods, products, instructions, ideas or otherwise contained in this publication.

This publication is designed to provide accurate and authoritative information with regard to the subject matter covered herein. It is sold with the clear understanding that the Publisher is not engaged in rendering legal or any other professional services. If legal or any other expert assistance is required, the services of a competent person should be sought. FROM A DECLARATION OF PARTICIPANTS JOINTLY ADOPTED BY A COMMITTEE OF THE AMERICAN BAR ASSOCIATION AND A COMMITTEE OF PUBLISHERS.

Additional color graphics may be available in the e-book version of this book.

Library of Congress Cataloging-in-Publication Data

ISBN: 978-1-53618-332-0
Names: Cowell, Richard I., editor.
Title: Dysbiosis : a study of underlying causes / Richard I. Cowell, editor.
Description: New York : Nova Science Publishers, [2020] | Series: New developments in medical research | Includes bibliographical references and index. |
Identifiers: LCCN 2020030173 (print) | LCCN 2020030174 (ebook) | ISBN 9781536183320 (paperback) | ISBN 9781536183580 (adobe pdf)
Subjects: LCSH: Human body--Microbiology. | Mouth--Diseases.
Classification: LCC QR46 .D97 2020 (print) | LCC QR46 (ebook) | DDC 616.9/041--dc23
LC record available at https://lccn.loc.gov/2020030173
LC ebook record available at https://lccn.loc.gov/2020030174

Published by Nova Science Publishers, Inc. † New York

CONTENTS

PREFACE

Dysbiosis: A Study of Underlying Causes includes studies aimed at determining design interventions in which the composition of microbiota may be modulated to achieve a specific microbial profile beneficial to overall host behavior.

In addition, this compilation summarizes and critically discusses recent literature on the evaluation of changes in the microbiota associated with liver diseases, aiming to evaluate which alterations could be involved in the initiation and progression of liver disease.

The authors emphasize the importance of promoting a balanced microbiome to maintain or improve oral health effectively. In order to fully understand the progression of oral diseases, it is essential to evaluate the microbiome together with genetic, immunological and environmental factors.

Chapter 1 - Gut-brain axis encompasses a complex communication network for maintaining metabolic homeostasis. These two systems regulate each other bi-directionally via multiple mechanisms and pathways, including neural, enteroendocrine, and immunological signals concerning mainly digestive function and satiety.

Recent studies proclaim microbiota as a new player in the gut-brain axis. Microbes communicate directly with the central nervous system and enteric nervous system via vagal and/or spinal afferents but also producing active

Richard I. Cowell

metabolites that influence enteroendocrine cells and gut mucosal immune system function. Other intermediates act indirectly crossing the intestinal barrier to enter the systemic circulation and passing the blood-brain barrier. In this context, altered microbiota can contribute to impaired metabolism through the gut-brain axis, and ultimately to the development of obesity.

Currently, obesity is a global epidemic. This disease has become a public health problem worldwide and has gradually increased during the last years. Evidence shows that the gut-brain axis and alterations in the gut microbiota are deeply related to obesity.

In the present review, the authors include studies aimed to design interventions in which the composition of the microbiota may be modulated to achieve a specific microbial profile beneficial to overall host behavior as well as exploring the link between the gut-brain-microbiota axis and its role in the development of obesity.

Chapter 2 - The liver is the central organ for metabolic processes including xenobiotic metabolism. An unhealthy status usually presents an altered gut microbiota profile (known as dysbiosis) associated with the loss of its functions.

The communication (gut-liver axis) between the liver and the gut is bidirectional and is mediated by the biliary tract, the portal vein, and the systemic circulation, with bile acids being key mediators in this communication. The influence of dysbiosis on the prevalence and pathogenesis of liver diseases is critical, ranging from simple steatosis (defined as excessive accumulation of triglycerides in hepatocytes) and steatohepatitis to more severe complications such as fibrosis, advanced cirrhosis and hepatocellular carcinoma. Additionally, taking into account the complex relationship among the gut microbiome, immune cells, and tumor cells in the liver, the characterization of alterations in the gut microbiota and their underlying causes related to the aforementioned diseases could have a great impact on the diagnosis, prevention, and design of potential therapeutic strategies. Knowledge in this regard could contribute to detaining the progression of disease from simple steatosis to hepatic cancer.

This review summarizes and critically discusses recent literature on the evaluation of changes in the microbiota associated with liver diseases. The authors aim to evaluate which alterations could be involved in the initiation and progression of liver disease.

Chapter 3 - The oral cavity, which has more than 700 bacterial species on teeth and mucosal surfaces, houses one of the most diverse microbial community in the human body. Distinct microenvironments containing heterogeneous microbes in the mouth form an important link for mouth and general health.

The microorganisms that make up the microbiome form highly regulated and well organized communities attached to surfaces as biofilms, which contribute to ecologic equilibrium. Substantial changes in a local environment can alter the competitiveness of biofilm bacteria may disturb this equilibrium, leading to enrichment of organisms most suited to the new environment, which can be defined as dysbiosis. The vast majority of oral diseases such as caries and gingivitis are caused by the ecological shift in oral biofilms.

Various genetic, epigenetic or patient-modifiable factors contributes to dysbiosis: such as immunological diseases, hormonal disorders, puberty, diabetes, stress, eating habits, smoking, antibiotic/antimicrobial agent use, poor oral hygiene and gingivitis cause changes in symbiotic bacterial community.

It is important to promote a balanced microbiome to maintain or improve oral health effectively. In order to fully understand the progression of oral diseases, it is essential to evaluate the microbiome together with genetic, immunological and environmental factors.

In: Dysbiosis: A Study of Underlying Causes   ISBN: 978-1-53618-332-0
Editor: Richard I. Cowell   © 2020 Nova Science Publishers, Inc.

Chapter 1

THE EFFECT OF GUT MICROBIOTA ALTERATIONS IN THE GUT-BRAIN AXIS AND ITS CONTRIBUTION TO THE OBESOGENIC STATE: AN OVERVIEW

Paula Robles-Bolívar[1] and Ana I. Álvarez-Mercado[2,3,4,*]
[1]Biochemistry and Molecular Biology I, School of Sciences,
University of Granada, Granada, Spain
[2]Department of Biochemistry and Molecular Biology II,
School of Pharmacy, University of Granada, Granada, Spain
[3]Institute of Nutrition and Food Technology "José Mataix,"
Biomedical Research Center, Granada, Spain
[4]Instituto de Investigación Biosanitaria ibs. GRANADA,
Granada, Spain

* Corresponding Author's Email: alvarezmercado@ugr.es.

ABSTRACT

Gut-brain axis encompasses a complex communication network for maintaining metabolic homeostasis. These two systems regulate each other bi-directionally via multiple mechanisms and pathways, including neural, enteroendocrine, and immunological signals concerning mainly digestive function and satiety.

Recent studies proclaim microbiota as a new player in the gut-brain axis. Microbes communicate directly with the central nervous system and enteric nervous system via vagal and/or spinal afferents but also producing active metabolites that influence enteroendocrine cells and gut mucosal immune system function. Other intermediates act indirectly crossing the intestinal barrier to enter the systemic circulation and passing the blood-brain barrier. In this context, altered microbiota can contribute to impaired metabolism through the gut-brain axis, and ultimately to the development of obesity.

Currently, obesity is a global epidemic. This disease has become a public health problem worldwide and has gradually increased during the last years. Evidence shows that the gut-brain axis and alterations in the gut microbiota are deeply related to obesity.

In the present review, we include studies aimed to design interventions in which the composition of the microbiota may be modulated to achieve a specific microbial profile beneficial to overall host behavior as well as exploring the link between the gut-brain-microbiota axis and its role in the development of obesity.

Keywords: gut-brain axis, gut microbiota, obesity, neuroendocrine system

INTRODUCTION

According to the World Health Organization, the main feature of obesity is the abnormal or excessive accumulation of fat with the concomitant health risk [1]. This disease and its related comorbidities have become a public health problem worldwide due to the great increment of the incidence in recent years [2].

Increasing evidence points that changes in gut microbiota composition and/or richness are deeply related to obesity. Indeed, obese humans and obese rodents have been shown to have a different microbial profile when compared with non-obese controls [3]. Alterations in the balance between the microbiota, intestinal permeability, and local immunity have an impact on homeostasis and negatively influence the entire organism [4]. Accordingly, the regulation of energy homeostasis guarantees an adequate balance between energy intake, storage, and expenditure to ensure the correct organism function. The gut-brain axis plays an important role in this task since both systems regulate each other bi-directionally via multiple mechanisms and pathways, including neural, enteroendocrine, and immunological networks. On the one hand, the central nervous system (CNS) integrates and processes continuous messages and stimuli about the energy state of the organism, developing an appropriate response for the maintenance of homeostasis. These stimuli are chemical and neuronal signals coming mainly from the gastrointestinal tract, but also from other viscera involved in the digestive process or metabolic balance, such as the pancreas or adipose tissue. Neural signals travel in the gut-brain axis through the three branches of the autonomic nervous system: 1) the sympathetic; 2) parasympathetic, and 3) enteric systems. Meanwhile, chemical signals (mostly hormones synthesized in the enteroendocrine cells of the gastrointestinal tract [GI]) act in an endocrine mode reaching the blood from the basolateral cell membrane, or in a paracrine way, stimulating receptors at the autonomous afferent endings. In addition to this interoceptive information, other areas within CNS modify eating behavior, such as the reward system and sensory and emotional aspects of feeding [5, 6].

Bearing in mind what we have mentioned so far, one of the possible mechanisms underlying the interplay between microbiota and host metabolism is through appetite-regulating hormones. This review aims to explore the connection between the gut-brain-microbiota axis and its possible role in either promoting or regulating energy by appetite-regulating hormones and their contributions to the obesogenic state.

NEUROENDOCRINE SYSTEM AND THE CONTROL OF ENERGY METABOLISM

Neuroendocrine System

The nervous and endocrine systems are intricately connected to each other and are responsible for the regulation of several physiological processes in the human body [7]. Together, they form the neuroendocrine system which is composed of the hypothalamus, pituitary gland, and target organs and involves one or more hypothalamic releasing or inhibiting hormones (known as hypothalamic axis, HPA) [8]. The hypothalamus controls the anterior pituitary systems regulating stress, basal metabolism, growth, reproduction and lactation. The released hormones from hypothalamic neurons terminate in the portal capillary vasculature where they are projected from the median eminence (ME) at the base of the hypothalamus to the anterior pituitary gland [7]. These hypothalamic neuroendocrine functions enable the central nervous system to respond rapidly to internal or external environmental changes and to maintain a response through endocrine hormonal transducers [7, 8].

Neuroendocrine cells are found in almost every organ of the body. We find them dispersed in the GI tract, the gallbladder, the pancreas and the thyroid. Neuroendocrine cells produce and release hormones and similar substances (peptides) in response to neurological or chemical signals. Next, the hormones enter the blood and travel throughout the body to the target cells and attach to specific receptors provoking changes [9].

This network is responsible for maintaining homeostasis and organizing the essential responses to inflammation or injury through a strictly regulated network of neuropeptides, hormones, cytokines, and chemokines [8].

Control of Energy and Metabolism

Digestive functions require an adequate integration of several systems working coordinately to maintain a homeostatic balance in the energetic state of the body. First, the gastrointestinal tract and accessory organs are responsible for sensing first line stimuli. Second, afferent vias convey inputs to high and low processing centers in the CNS, where information will be processed and integrated. Finally, elaborated responses are transmitted to the specialized motor effector system to be executed [6].

Brain central circuits must integrate homeostatic-metabolic interoceptive information comprising endocrine and neural signals with non-homeostatic exteroceptive-hedonic inputs, and ultimately to assign a value for food and determining the eating behavior [10]. Homeostatic-metabolic regulation is performed by the hypothalamic arcuate nucleus (ARC), that contains two specialized neuronal types, with opposite activities in terms of appetite and satiety provocation: orexigenic agouti-related peptide and neuropeptide Y expressing (AgRP/NPY) neurons and anorexigenic neurons proopiomelanocortin expressing (POMP) neurons [11, 12]. Moreover, ARC possesses a privileged position near the ME, where exists a blood-brain-barrier leaky region allowing the entry of peripheral enteroendocrine signals such as hormones and nutrients [13]. Both AgRP/NPY and POMC second-order neurons project downstream to paraventral nucleus (PVN) and lateral hypothalamic area (LHA). PVN is associated with intake inhibition, containing anorectic factors such as corticotropin-releasing hormone (CRH) and thyrotropin-releasing hormone (TRH), carrying out a direct regulation function on the HPA. Conversely, LHA is associated with stimulation of intake, containing orexigenic neuropeptide factors such as hypocretin/orexin, melanin-concentrating hormone (MCH), neurotensin and histamine [14]. In parallel, several cortico-limbic regions integrate non-homeostatic information including sensory (taste, smell), motivational (reward system), emotional and memory-learning aspects of feeding [10, 15].

Effective communication between brain centers and enteric nervous system is mediated by both sympathetic and parasympathetic branches of autonomic nervous system through "top-down" (brain to visceral) efferent motor neurons and "bottom-up" (visceral to the brain) afferent pathways [16].

To date, the interaction between gut microbiota and neuroendocrine axes has not been elucidated even though it is known that the gut microbiota affects several physiological and behavioral outcomes through modulation of neuroendocrine pathways [3].

HORMONES AND NEUROPEPTIDES AS GASTROINTESTINAL ENERGY MEDIATORS

The control of energy metabolism guarantees an adequate balance between energy intake, storage, and catabolism to ensure the correct organism functioning, without the development of pathologies such as obesity [17]. The gut-brain axis is a key player in this task since both systems interplay by numerous mechanisms and pathways, including neural, neuroendocrine, and immunological networks [5]. Although the precise mechanisms through the gut microbiota might influence neuroendocrine functions have not yet been fully deciphered [3], two possible scenarios are identified: a) a prolonged fasting situation where reserves decrease and appetite must be induced to cause a short-term intake and a decrease in energy expenditure; b) after ingestion and, with sufficient reserves, satiety must be promoted and basal metabolism increased [18].

Evidence suggests that the gut microbiota acts through direct production of neuroendocrine metabolites (hormone-like metabolites such as short-chain fatty acids (SCFAs), neurotransmitters, GI hormones and precursors to neuroactive compounds such as tryptophan and kynurenine) and, indirectly, as modulator of inflammatory and immune responses as well as hormonal secretion. The most immediate mechanism through which the gut microbiota modulates food intake is through the production/alteration of appetite-regulating hormones [3].

The signals to decrease appetite and increase energy expenditure are called anorexigenic and are released from many systems such gastrointestinal tract: peptide YY (PYY), glucagon-like peptide (GLP-1), cholecystokinin (CCK), oxyntomodulin, obestatin, apolipoprotein A-IV, a part of gastrin-releasing peptide (GRP), neuromedin B, neurotensin, enterostatin, serotonin and glucose; pancreas: pancreatic polypeptide (PP), insulin, glucagon and amylin; adrenal gland: adrenalin and noradrenalin (β-adrenergic effects) and oestrogens, although another fraction of oestrogens came from ovary and placenta; adipose tissue: leptin, omentin and visfatin; hypothalamus: α-melanocyte-stimulating hormone, CRH, TRH, adrenocorticotropic hormone (ACTH), cocaine and amphetamine-regulated transcript peptide, oxytocin and nesfatin; and several brain regions in peripheral nervous system (PNS) and CNS: brain-derived neurotrophic factor and another fraction of serotonin and gastrin-releasing peptide. In turn, stimuli to increase appetite and decrease energy expenditure are called orexigenic and come from the gastrointestinal tract: ghrelin, motilin and some galanin; adrenal gland: glucocorticoids, adrenalin and noradrenalin (α-adrenergic effects) and a fraction of androgens and progesterone; testicles and ovaries with another androgens and progesterone part; pituitary gland with growth-hormone (GH); hypothalamus: neuropeptide Y (NPY), agouti-related peptide (AgRP), MCH, growth hormone-releasing hormone (GHRH), orexin/hypocretin, somatostatin or somatotropin release inhibiting hormone, cerebellin1 and another part of galanin; and CNS: glutamate, gamma-aminobutyric acid and endocannabinoids [10, 14, 18-22]. The most deeply associated with energy, appetite and obesity are summarized in Table 1.

Table 1. Gut hormones and neuropeptides involved in appetite an energy regulation. Sites of secretion, major functions and alterations in obesity

Hormone	Site of secretion	Target receptors and/or location	Major actions	Refs.	Obesity	Refs.
			DECREASE APPETITE AND INCREASE ENERGY EXPENDITURE			
GLP-1	EECs L cells (GIT)	GLP-1R (pancreas, GIT, vagal system, NTS, hypothalamus)	Increase insulin synthesis and secretion and inhibit glucagon release in pancreas Delays gastric emptying and slow GIT motility Low gastric acid secretion Enhance insulin sensitivity and glucose uptake in muscle Decrease hepatic gluconeogenesis Induce satiety activating POMC and inhibiting AgPR/NPY neurons in hypothalamus Reduce food intake and reduce weight gain Decrease reward eating behaviour	[23-26]	Decreased GLP-1 plasma postprandial levels in overweight/obese individuals Polymorphism in GLP-1R is associated with BMI GLP-1 infusion diminished food intake dose-dependently in obese and lean subject Potential target for anti-obesity therapy	[27-31]
PYY	EECs L cells (distal jejunum and ileum)	Y2R (vagal system, hypothalamus, GIT)	Low gastric acid secretion Delays gastric emptying and slow GIT motility Stimulate glucose-induced insulin release Reduce food intake and reduce weight gain	[32-34]	Lower PYY basal fasting levels and smaller PYY rise in postprandial secretion in obese compared to lean subjects Absence of PYY resistance PYY infusion in obese subject reduce food intake	[35-37]

Hormone	Site of secretion	Target receptors and/or location	Major actions	Refs.	Obesity	Refs.
DECREASE APPETITE AND INCREASE ENERGY EXPENDITURE						
			Induce satiety activating POMC and inhibiting AgPR/NPY neurons in hypothalamus		Potential target for anti-obesity therapy	
CCK	EECs I cells (duodenum and jejunum)	CCK1R, CCK2R (GIT, pancreas, gallbladder, vagal system, hypothalamus, NTS, CNS)	Stimulate gallbladder contraction Accelerate gastrointestinal motility Rise pancreatic enzyme secretion but low gastric acid secretion from stomach Increase insulin but inhibit glucagon secretion Delays gastric emptying Induce satiety via CCKR in vagal afferent and hypothalamus Reduce food intake	[38–41]	Unclear CCK levels alteration in obesity CCKR polymorphism associated with diabetes and obesity development Vagus nerve insensitivity to CCK	[21, 35, 42, 43]
Leptin	Adipose tissue	Ob-Rb (hypothalamus, vagal system, periphery tissues)	Long-term regulation of energy balance Increase energy expenditure and reduce body weight gain and food intake Induce satiety activating POMC and inhibiting AgPR/NPY and GABA neurons in hypothalamus Decrease adiposity, stimulating WAT browning and BAT thermogenesis Stimulate insulin and GH secretion	[44–47]	Central and peripheral leptin resistance Increased leptin plasma levels and leptin mRNA expression in obese compared to lean subjects Associated hyperinsulinemia and insulin resistance Leptin and leptin receptor gene mutations associated with early obesity development	[48–51]

Table 1. (Continued)

Hormone	Site of secretion	Target receptors and/or location	Major actions	Refs.	Obesity	Refs.
DECREASE APPETITE AND INCREASE ENERGY EXPENDITURE						
Omentin	Adipose tissue	Adipocytes	Enhance insulin sensitivity Increase insulin-dependent glucose uptake in adipocytes Anti-inflammatory, anti-atherosclerotic and cardiovascular protective effects	[52-54]	Decreased omentin plasma levels and omentin mRNA expression in obese compared to lean individuals Decreased omentin levels associated with insulin resistance Plasma omentin levels correlate negatively with BMI, waist circumference, leptin levels and insulin resistance	[55, 56]
Visfatin	Adipose tissue	IR	Intervene in glucose homeostasis regulation Exert an insulin mimetic activity Low plasma glucose levels Increase insulin-dependent glucose uptake in adipocytes	[57, 58]	Visfatin overproduction by visceral adipose tissue Increased visfatin levels positively correlated with BMI, inflammation, angiogenesis, insulin resistance and pancreatic β-cell dysfunction in obesity Potential target for anti-obesity therapy	[59-61]
Obestatin	Gastric cells, pancreas, adipose tissue	GLP-1R (pancreatic β-cells, GIT, NTS, vagal system, hypothalamus)	Delays gastric emptying and slow GIT motility Increase pancreatic enzymes secretion Inhibit glucose-induced insulin secretion	[62-64]	Decreased obestatin plasma levels in obese compared to lean subjects. Obestatin levels negatively correlated with body mass index (BMI), leptin, glucose and insulin	[65-67]

Hormone	Site of secretion	Target receptors and/or location	Major actions	Refs.	Obesity	Refs.
			DECREASE APPETITE AND INCREASE ENERGY EXPENDITURE			
			Increase adipogenesis and glucose and fatty acids uptake in adipocytes Reduce food intake and body weight gain		but positively associated with ghrelin Obesity obestatin related to insulin resistance	
PP	Pancreatic PP cells	Y5R (pancreas and gallbladder) Y4R (Hypothalamic ARC and PVN, NTS) Y1R (vagal system)	Stimulate gallbladder contraction Decrease exocrine secretion in pancreas Delaying gastric emptying Decrease leptin and ghrelin levels Increase basal metabolism and energy expenditure Reduce food intake and body weight Stimulate anorexigenic and decrease orexigenic peptides secretion from hypothalamic areas Activate α-MSH signalling pathways Stimulate vagal activity and sympathetic reflex	[68, 71]	Lower PP increased in postprandial level in obese compared to lean subjects Reduced PP sensitivity PP levels normalize after weight loss PP administration decrease body weight and ameliorated insulin resistance and hyperlipidaemia in obese mice Potential target for anti-obesity therapy	[71-73]
Insulin	Pancreatic β-cells	IR (hypothalamus, liver, muscle, adipocytes)	Intervene in long-term regulation of energy balance and glucose metabolism Decrease plasma glucose levels, suppresses gluconeogenesis and increases glycogenesis in liver	[44, 74-76]	Central and peripheral insulin resistance Increased fasting insulin levels in obese compared to lean individuals	[44, 76, 78]

Table 1. (Continued)

Hormone	Site of secretion	Target receptors and/or location	Major actions	Refs.	Obesity	Refs.
			DECREASE APPETITE AND INCREASE ENERGY EXPENDITURE			
Insulin (continued)			Increase glucose uptake in adipose tissue and muscle		Visceral fat, saturated fatty acids and BBB alterations related to insulin resistance development	
			Suppresses lipolysis, induce WAT browning and stimulate FFA uptake in adipose tissue		Low glucose oxidative metabolism	
			Reduce energy intake and body weight and increase energy expenditure		Unrestrained lipolysis and consequent lipotoxicity	
			Induce satiety activating POMC and inhibiting AgPR/NPY and GABA neurons in hypothalamus		Insulin gene polymorphism associated with obesity development	
			Improve cognitive, memory and reward aspects		Overcoming insulin resistance as anti-obesity therapy	
					Insulin sensitivity normalize after weight loss	
Amylin	Pancreatic β-cells	AMY (Hypothalamus, NTS)	Intervene in long-term energy balance regulation	[79-81]	Increased amylin plasma levels in obese compar to lean	[79, 81, 82]
			Inhibits gastric secretion		Down-regulated of amylin receptors	
			Delays gastric emptying		Reduced satiety provocation	
			Inhibit insulin and glucagon secretion		Amylin levels positively associated with insulin, LDL–cholesterol and triglycerides concentration in obese children compared to lean	
			Reduce food intake and body weight			
			Induce satiety inhibiting NPY release from NPY/AgRP neurons on hypothalamic AP		Brain infusion of amylin reduce body weight gain and adiposity,	

Hormone	Site of secretion	Target receptors and/or location	Major actions	Refs.	Obesity	Refs.
					independently of starting body weigh Potential use of amylin and leptin as anti-obesity treatment	
DECREASE APPETITE AND INCREASE ENERGY EXPENDITURE						
Glucagon	Pancreatic α-cells	GCGR	Enhancing physiological response to stress Increase plasma glucose levels activating gluconeogenesis, glycogenesis and lipolysis Reduce body weight and adiposity by suppression of appetite and modulation of lipid metabolism Promote weight loss by activation of energy expenditure and thermogenesis	[83-86]	Obese hyperglucagonemia: increased glucagon plasma levels compared to lean Exacerbate hyperlycemia Glucagon levels possitively correlated with hiperinsulinemia, visceral adiposity, high plasma FFAs and triglycerides. Predispose to T2D development Static glucagon:insulin ratio in obese compared to lean individuals in fasting state. Possible potential anti-obesity therapy	[87, 88]
INCREASE APPETITE AND DECREASE ENERGY EXPENDITURE						
Ghrelin	X/A-like cells (oxyntic gastic gland, stomach, hypothalamus)	GHS-R (peripheral tissues, vagal system, ARC)	Accelerate gastric empyting and GIT motility Lower meal-related glycemia Inhibit insulin secretion Stimulate appetite, activating AgRP/NPY neurons and inhibiting POMP neurons in ARC	[89-92]	Decreased ghrelin plasma levels in obese compared to lean subjects ARC ghrelin resistance: reduced NPY/AgRP sensitivity to ghrelin and decreased GHS-R Ghrelin regulation of vagus nerve plasticity toward fasting state	[30, 35, 50, 93, 94]

Table 1. (Continued)

Hormone	Site of secretion	Target receptors and/or location	Major actions	Refs.	Obesity	Refs.
INCREASE APPETITE AND DECREASE ENERGY EXPENDITURE						
Ghrelin (continued)			Block leptin-induced satiety Stimulate HPA axis and GH secretion Promote fat accumulation and body weight gain		GHR polymorphism association with BMI variation in human	
GHRH	GHRH neurons (Hipothalamic ARC)	GPCR (pituitary gland somatothrops)	Stimulate growth hormone (GH) secretion from pituitary gland	[95, 96]	Decreased GHRH expresion and secretion Somatothrops GHRH insensitivity Decreased GHRH mRNA content and expression in obese Zucker rats GHRH treatment reverse to normal secretion and frequency GH pulse, also with decrement in visceral fat, dislipemia and cardivascular tisk marker	[47, 97, 98]
SRIH	Somatostatin neurons (Hypothalamic PVN)	SHRI receptors (CNS and periphery)	Rise gastric acid secretion and propulsive motor function in GIT Increase blood glucose levels Inhibit growth hormone (GH) releasing Downstream stimulate NPY neurons	[95, 99, 100]	Unclear somatostatine release alteration in obesity SHRI levels correlated negatively con BMI in obese patient SHRI and analogues reduce fat content	[101, 102]

Hormone	Site of secretion	Target receptors and/or location	Major actions	Refs.	Obesity	Refs.
			INCREASE APPETITE AND DECREASE ENERGY EXPENDITURE			
			Increase food and water intake and energy expenditure Promote body weight loss			
GH	Pituitary gland	GHR (peripheral tissues, ARC)	Stimulate hepatic IGF secretion Increase hepatic gluconeogenesis, glycogenolysis and TG secretion Induce lipolysis in adipose tissue Increase lipid uptake and mobilization in skeletal muscle Stimulate appetite	[96, 103, 104]	Decreased GH secretion: reduced half-life, frequency and secretory episodes Reduced GHR expression in adipose tissue Exacerbate obesity fat accumulation No response to GH stimulators: insulin, glucocorticoids, hypoglycaemia, GHRH... Higher free IGF-1 aggravate GH blunted state	[47, 96, 105, 106]
GC	Adreno-cortical cells (adrenal gland)	GR (peripheral tissues, ARC)	Glucose mobilization in stress response Stimulate liver gluconeogenesis and glycolysis Adipocytes lipolysis stimulation Peripheral proteolysis stimulation (muscle) Modulate GIT peptides release decreasing PYY and ghrelin secretion Increase pancreatic insulin and glucagon	[107, 108]	Increased obesity glucocorticoid levels in tissues Hyperactivate HPA activity High GC correlated with hyperglycaemia, hypertriglyceridemia, dislipemia, high leptin and insulin resistance Chronically raised GC levels supress GH secretion from pituitary gland	[108-111]

Table 1. (Continued)

Hormone	Site of secretion	Target receptors and/or location	Major actions	Refs.	Obesity	Refs.
INCREASE APPETITE AND DECREASE ENERGY EXPENDITURE						
GC (continued)			secretion			
			Stimulate NPY and decrease CRH, ACTH and POMP expression or release			
			Induce food intake and body weight gain			

Abbreviations: AgRP: agouti-related protein; AMY: amylin receptor; AP: area postrema; ARC: arcuate nucleus; BAT: brown adipose tissue; BBB: blood brain barrier; BMI: body mass index; CCK: cholecystokinin; CCK1/2R: cholecystokinin 1/2 receptor; CNS: central nervous system; EECs: enteroendocrine cells; FFA: free fatty acids; GABA: gamma aminobutyric acid; GC: glucocorticoid; GCGR: glucagon receptor; GH: growth hormone; GHR: growth hormone receptor; GHRH: growth hormone-releasing hormone; GHIH: growth hormone-inhibiting hormone (somatostatin) GHS-R: growth hormone secretagogue receptor; GIT: gastrointestinal tract; GLP-1: glucagon-like peptide 1; GLP-1R: glucagon-like peptide 1 receptor; GPCR: G protein-coupled receptor; GR: glucocorticoid receptor; HPA: hypothalamic-pituitary-adrenal; IR: insulin receptors; NPY: neuropeptide Y; NTS: nucleus tractus solitarius; Ob-R: leptin receptor; POMP: propiomelanocortine; PP: pancreatic polypeptide; PVN paraventral nucleus; PYY: polypeptide YY; Y1/2/4/5R: neuropeptide Y1,2,4,5 receptor; WAT: white adipose tissue.

RECENT STUDIES ABOUT THE RELATIONSHIP BETWEEN THE NEUROENDOCRINE SYSTEM, THE GUT MICROBIOTA AND OBESITY

The gut microbiota affects several physiological and behavioral outcomes (including obesity) through modulation of neuroendocrine pathways. People with obesity manifest deficits in memory and learning, as well as high risk of depression and anxiety development. Associations between stress and the onset of obesity are prevalent in Westernized societies and have been linked with intestinal microbiota derangement. Some investigations illustrated the relationship between greater amount of abdominal fat and over-excitation and imbalance in the HPA axis, the main regulator of the stress response, finding changes in neuroendocrine and monoaminergic metabolites, such increase in cortisol and glucocorticoid levels and abrupt inhibition of growth hormone release, as well as microbiota alterations and deregulating vagal signaling and immune mediators [112, 151].

However, to date, the interaction between gut microbiota and the gut-brain axis has not been completely elucidated.

Probiotics, prebiotics, as well as complex fibers, oligosaccharides or plant bioactive compounds have been used as strategy to restore optimal gut bacterial composition (eubiosis). Nonetheless, weight loss and bariatric surgery approaches continue being the most effective therapies recovering obesity symptoms such as the reduction of exacerbated appetite, weight and fat content, low-grade inflammation reversion, restoration of the gastrointestinal hormones secretion and the return of the gut microbiota to eubiosis [113].

In this section, we include reported studies showing interventions and/or pathways in which the composition of the microbiota is altered by changes in which the hormones and neuropeptides showed in table 1 are involved. These studies, covering the last 5 years, are summarized in table 2. Remarkably, many hormones and neuropeptides related to the control of energy metabolism and balance between energy intake, storage, and catabolism did not display any related-study (i.e., visfatin, obestatin, pancretatic peptide, amylin, glucagon, GHRH and somatostatin).

PYY and GLP-1

GLP-1 is secreted in response to meal by enteroendocrine L cells, located throughout the entire intestine. Their function is triggered by the interaction with GLP-1 receptors (GLP-1R), expressed in the gastrointestinal tract, pancreas, vagal afferents, and many brain areas such as the hypothalamus and nucleus tractus solitarius. GLP-1 is an anorexigenic peptide, which induces a decrease in appetite and, consequently, in food intake at the central level, mainly acting in the ARC, by inhibition of AgRP/NPY and activation of POMC neurons [24, 26]. However, neuronal excitation with GLP-1 have also been observed in areas such as ventral tegmental area, amygdala or nucleus accumbens, relating it to reward, memories and learning aspects of food [25]. At metabolic level, GLP-1 modulates glucose homeostasis, affecting the regulation of insulin and glucagon release, inhibiting hepatic glucose synthesis, increasing insulin sensitivity and glucose uptake in muscle, delaying gastric emptying and decreasing gastric acid secretion and motility in the stomach [23, 26]. Meal-induced GLP-1 secretion has been revealed as down-regulated in obese people [27-29]. Increased body mass index (BMI) has also been associated with GLP-1 receptor polymorphisms, suggesting that defects in signaling may promote disease development [30]. Data from multiple clinical trials where GLP-1 was administered to obese patients have been reviewed, resulting in a decrease in appetite and food intake in a GLP-1 dose-dependent manner, indicating a potential therapeutic role [31].

Enteroendocrine L cells in the distal jejunum and ileum also co-synthesize and release PYY. PYY is secreted in proportion to caloric intake. Once released, part of PYY is enzymatically cleaved to major circulating form PYY_{3-36}, which carries out its function through Y2 receptors, widely expressed in the gastrointestinal tract, vagal afferents and several central regions such as the hypothalamic ARC. Similar to GLP-1, PYY anorexigenic effects encompass meal-related glycaemia reduction, glucose-induced insulin release, glucose uptake in muscle and adipose tissue, acid gastric secretion inhibition and gastric emptying delays slowing gastrointestinal motility. PYY stimulates satiety and energy expenditure centrally, acting on hypothalamic ARC neurons [32-34]. Fasting baseline PYY levels in obese subjects are lower if compare with healthy subjects, being also smaller the post-prandial PYY expected increment [35, 36]. However, it seems that there is not peptide resistance since infusions of PYY in obese subjects produce intake reductions of up to 30% [36, 37].

As indicated, the secretion of PYY and GLP-1 is decreased in obesity. Various strategies focused on microbiota modulation have demonstrated effectiveness in reversing this anomaly in animal models and human patients, increasing the levels of these gastrointestinal hormones. Changes in the composition of the microbiota have also been reported. Regarding probiotics, administration of *B. animalis spp. Lactis* GLC2505 (BlaG) and *B. longum* JCM1217T (BloJ) in Aoki et al. research, and *Lactobacillus rhamnosus, L. acidophilus* and *Bifidobacterium bifidumi* by Bagarolli et al. group in mice with HFD-induced obesity stimulated induction in GLP-1 secretion, in addition to other metabolic improvements. Fecal microbiota analysis revealed enrichment in *Bifidobacterium,* [114] continued presence of *Bacteroidetes* and *Actinobacteria* and a decrease in *Firmicutes* [115]. In a clinical assay with metabolic syndrome subjects, the intake of 67 probiotic strains in association with prebiotic fructooligosacaride also increased GLP-1 and PYY, accompanied by improved insulin response and reduced BMI [116].

Other strategies have also tackled the role of prebiotic slow-rate metabolizing fiber, such as the inclusion of barley β-glucan and inulin in the diet of obese HDF mice and rats, both showed an increase in GLP-1 and PYY. Gut microbiota profile in response to β-glucan fermentation revealed a similar *Firmicutes:Bacteroidetes* ratio that non-treated animals, but amplification in the main acetate producer *Actinobacteria*. This was in accordance with an increase in SCFAs levels [117]. On their behalf, Singh's group evidenced that inulin rise *Bacteroidetes* and decreased *Clostridium* [118]. Supporting this, inulin propionate-ester conjugation also produced an increase in GLP-1 and PYY secretion in obese adults, in addition to other metabolic improvements such as lower abdominal adiposity. Fecal content presented an enlarged *Bacteroides* and *Atopobium* population [119].

Less complex carbohydrates compounds that appear in advanced stages of food decomposition during digestion have also been tested since these are fermented and modified by the microbiota at colonic level. For example, obese HFD rat diet supplemented with galactooligosaccharides (GOS), oligofructose (OFS) or epigallocatechin-3-gallate (EGCG) have been tested in rats. The results showed that animals presented increased levels of GLP-1 and PYY in addition to microbial profiles remodeling, and increased *Akkermansia muciniphila* in response to EGCG [120]. In contrast, GOS favored a considerable growth of *Bifidobacteria* [121], and conversely a decrease in *Akkermansia muciniphila* caused by OFS [116]. The increase in GLP-1 was also verified with chondroitin sulphate, a glycosaminoglycan usually associated to protein in animal tissues, and substrate for sulphate-reducing bacteria, which generated *Desulfovibrio piger* proliferation and H_2S in fecal bacterial composition [122].

Many plant components have dietary properties and are usually used in weight loss purposes. For instance, berberine dispensed to obese HDF rats caused an increase in hypothalamic levels of GLP-1R, along with weight and fat content loss. Although no significant differences were found at phyla levels, genus shift was observed with *Bacteroides, Sutterella, Bilophila, Desulfovibrio* and *Aenaerofilum* enrichment and *Blautia, Dorea* and *Roseburia* decrement [123]. Capsaicin administered to ob/ob mice enhanced GLP-1 and decreased inflammatory markers together with an augmentation of *Firmicutes:Bacteroidetes* ratio, expansion of *Roseburia* and reduction of *Bacteroides* and *Parabacteroidetes* abundances [124].

Various studies in diet-induced obese rats have focused on analyze the changes produced after popular Roux-en-Y bypass and vertical sleeve gastrectomy, introducing new approaches such as one-anastomosis gastric bypass or single-anastomosis duodenal switch. All these studies revealed induced decrease in weight and abdominal fat, together with the improvement of lipid and glucose homeostasis, enhancing of insulin sensitivity and an increase of PYY and GLP-1 levels. Also, microbiota composition reverted from obesogenic state in those procedures, although this shift was not validated in sleeve gastrectomy in the study reported by Arble [125, 126]. These results were corroborated in severely obese humans submitted to biliointestinal bypass, along with increases in *Lactobacillus crispatus* and *Meghasphaera elsdenii* and decreased *Clostridial XIVa* and I*V* clusters [127].

Cholecystokinin (CCK)

Cholecystokinin (CCK) is an anorexigenic peptide synthesized in enteroendocrine I cells mainly in the duodenum and jejunum. CCK carries out its action by binding to CCK1/2R receptors, found in a wide variety of peripheral and central tissues. Among its metabolic functions, it stimulates bladder contraction, gastrointestinal motility, and pancreatic enzyme secretion, and delays gastric emptying [38, 41]. Moreover, CCK induces a feeling of satiety, reducing intake and increasing energy expenditure, thought activation of CCK receptor in vagal afferents and thus signaling to CNS [39, 40]. The role of CCK in obesity is unclear. Some studies have found CCK overproduction while others exposed decreasing levels in obese compared to healthy subjects [21, 35]. However, polymorphisms or mutations in CCK receptors leading to disease have been described in obese patients [42]. Furthermore, vagal plasticity can turn towards an "obese phenotype" exhibiting insensitivity to the action of the hormone in vagal afferents [43].

If prebiotic or probiotic can improve response to CCK signaling and restore obesity-associated microbiota alteration has been examined. For example, 2'-fucosyllactose, the third most abundant oligosaccharide in breast milk, which has been related to the correct development of gut microbiota was administered to HFD mice. The treatment improved CCK sensitivity in animals in addition to expand *Parabacteroides* [128]. A similar improvement in CCK sensitivity was also observed in rats supplied with potato resistant starch [129]. Other strategies included increasing dietary protein also improved CCK sensitivity and *Akkermansia muciniphila* enlargement [130]. Regarding vegetal compounds, thilakoids treatment showed modulation of the microbiota in obese human and obese rats, with an increase in *Bacteroides fragilis*, but without response in CCK levels. These investigations were accompanied by improvement in anthropometric parameters, with a decrease in body fat, weight, hyperphagia and systemic inflammation [131].

Ghrelin

Ghrelin is a hormone secreted in response to fasting. It is mainly synthesized by the oxyntic glands in the stomach, although another part is produced in the small gut and the CNS. Ghrelin acts through the growth hormone secretagogue receptor (GHS-R), expressed in both peripheral and central tissues, as well as in vagal afferents [89]. At central level, ghrelin produces an orexigenic effect, stimulating the appetite in ARC through activation of AgRP/NPY and inhibition of POMC neurons [91]. Many of the downstream effects of ghrelin on the hypothalamus are related to activation of the HPA axis and GH secretion, promoting increased intake and weight, modulating meal-induced glycaemia, inhibiting leptin-induced feeding reduction and insulin release, as well as accelerating gastric emptying and stimulating gastric motility [90, 92]. A large number of studies have found decreased ghrelin levels in obese subjects [35, 50]. Anyway, available ghrelin fails its function because arcuate neurons and vagal afferents develop ghrelin resistance, characterized by a lack of sensitivity to the hormone along with a decrease in the expression GHS-R [93, 94]. GHS-R receptor polymorphisms have been also associated with increased BMI [30].

Diet enrichment with some products to combat obesity have evidenced conflicting results according to ghrelin. Exogenous acetate carrier (acetate bound to resistant starch), and plant components such as cinnamaldehyde or capsaicin have been administered to HFD rats or ob/ob mice in feeding. In all of them, anti-obesogenic functions were observed, preventing weight gain and hyperphagia, improving insulin sensitivity and systemic inflammation, or enhancing lipolysis. Concerning microbiota, the *Firmicutes:Bacteroidetes* ratio decreased, and some major butyrate producers such as *Roseburia, Bacteroides, Parabacteroidetes, Eubacterium, Coprococcus* and *Faecalibacterium* augmented. However, acetate and capsaicin produced lower ghrelin levels [124, 132], while cinnamaldehyde increased it [133]. The same contradictory outcome has been also observed after surgeries. Ghrelin decreased in HFD rats submitted to glandular gastrectomy accompanied by enlarged *Lactobacillus* and *Collinsella* populations [126]. However, bariatric surgery augmented ghrelin levels in severely obese humans and turned gut community from mainly *Clostridial XIV* and *IV* clusters before surgery to *Lactobacillus crispatus* and *Meghasphaera elsdenii* after [127].

Leptin

Leptin is an adipocyte-derived hormone that regulates satiety, energy expenditure, immune system, hematopoiesis, angiogenesis, reproduction and carcinogenesis [134].

Both, leptin and insulin act to reduce food intake and increase energy expenditure via action on hypothalamic neurons in a process known as "body adiposity signals"[135]. Leptin crosses the blood-brain barrier, acts on the hypothalamus and regulates energy metabolism by appetite decrease and increasing energy expenditure with heightened sympathetic system activity [136].

Liver and stomach are leptin producers [134, 137]. Higher expression of gastric leptin and its receptor signaling results in gastric malignancy (i.e., cancer) [137]. Besides, it is thought that obese individuals present complications such as resistance to the hormone and alteration of the receptors in the hypothalamus, or transport [18].

Concerning studies covering the last 5 years focused on decipher the interplay between leptin and microbiota in the context of obesity, the majority were conducted in leptin-deficient ob/ob mice and rodent models of obesity-induced by high fat diet (HFD) (Table 2). Leptin-deficient ob/ob mice have single-gene loss-of-function mutations regarding leptin metabolism that generates massive obesity [138] which allow to selectively examine the precise function of leptin. Interventions with probiotics [115, 139], plant-derived supplements [140-142] or surgical procedure [143] contributed to ameliorate insulin and leptin resistance [115, 142], and reversion of obesity-related features [115, 141-143]. Besides, strategies aimed to increase circulating leptin levels, provoked increments in the expression of fat hydrolysis and oxidative genes [144] by modulations of leptin signaling and gastric microbiota [145]. Also, leptin showed a modulatory role of gut microbiota on the development of intestinal functionality [146]. Additionally, under leptin-deficient conditions the gut structure was particularly altered [147].

In this line, studies in humans have linked serum leptin to gut microbiota suggesting that leptin itself may regulate gut microbiota. In breast milk-derived, gut-seeding gut *Enterococci* correlates inversely with excessive weight gain and increased plasma leptin [148]. In addition, alterations in leptin and insulin were related to early changes in the infant gut microbiome. Both hormones were independently associated with microbial metabolic pathways associated with an increase in intestinal barrier function and decreased intestinal inflammation [149].

Omentin-1

The omentin-1 is secreted from the adipose tissue, mainly from the visceral fat [150]. There is an inverse correlation between omentin -1 levels and obesity [150] and is considered protective against carcinogenesis and inhibitor of vascular endothelial cells inflammation [134]. Firstly, it was thought that Omentin-1 was part of the early-defense mechanisms against pathogenic bacteria in the gut [151]. A study reported by Sanchis-Chorda et al. found increased omentin-1 levels in children supplemented with the probiotic *B. pseudocatenulatum* CECT 7765 when compared to the control group. Although the changes in gut microbiota were minor and there was not found a direct correlation with omentin-1 [152].

Insulin

Insulin is secreted by the β cells of the pancreatic islets of Langerhans. Insulin receptors are expressed in the mediobasal hypothalamus, and ME [153].

Its main role is to maintain blood glucose levels by facilitating cellular glucose uptake, regulating carbohydrate, lipid and protein metabolism and promoting cell division and growth through its mitogenic effects [154]. Levels of plasma insulin vary directly with changes in adiposity so that plasma insulin increases at times of positive energy balance and decreases at times of negative energy balance [153].

Several studies point that the gut microbiota can affect the physiology of the host and may thereby contribute to insulin resistance. Similarly to the strategies addressed to leptin modulation, interventions in both humans and rodent model with probiotics [155] plant-derived supplements [156-158], and surgical procedures [158-162] have been widely used to improve both, obesity and metabolic syndrome thought the regulation of insulin levels or the amelioration of insulin resistance. Also, antibiotics have been tested. For instance, the treatment with amoxicillin or vancomycin for 7 days had not any effect in insulin sensitivity in obese men with impaired glucose homeostasis [163]. However, the majority of the studies aforementioned revealed anti-obesogenic, anti-inflammatory and antidiabetic effects connecting gut microbiota richness, diabetes risk in overweight/obese humans, and insulin (Table 2).

Growth Hormone

GH is released from the anterior pituitary gland [164]. Besides its effects on growth, GH deeply affects energy metabolism. Thus, reduced GH secretion provokes decreased fat metabolism, decreased muscle mass, and increased fat stores, all together contribute to the development of obesity and cardiovascular disease. Accordingly, obesity is known to cause a decrease in GH secretion [165]. The relation of GH with gut microbiota has not been examined in detail. From our knowledge, only one study has been reported in the last 5 years suggesting that GH could promote the growth of some bacteria implied in the microbial maturation and metabolic functions (i.e SCFA, folate, and heme B biosynthesis) [166].

Glucocorticoid

Glucocorticoid release is induced as a result of HPA axis activation in response to stress to ensure the presence of an easy-to-use energy source. This response includes glucose mobilization, hepatic de novo glucose synthesis, adipose tissue lipolysis and proteolysis in skeletal muscle [107]. Persistent activation of the HPA axis has been observed in obese individuals, leading to an elevated endocrine response. In fact, obese subjects usually present elevated glucocorticoid levels [109], [108] and GC receptors decreased [139, 167]. The use of glucocorticoid as treatments for obesity worsens this issue, reinforce the over-activation of the axis and accentuate the alteration of the microbiota even more than the changes produced by obesity itself. This leads to an enrichment of *Firmicutes* and depletion of *Bacteroidetes,* along with a decrease in SCFAs production [168]. High GC levels have been associated with impaired metabolic features such as high leptin and insulin resistance, hyperglycemia, hypertriglyceridemia and dyslipidemia [108]. Concerning the role of *Bifidobacterium pseudocatenulatum* in the HPA axis beneficial effects have been demonstrated in obese rats fed with HFD. These animals presented depressive behavior amelioration, reduced corticosterone levels and increased glucocorticoid receptor expression at the hippocampus level [139].

Table 2. Studies about the relationship between the neuroendocrine system, gut microbiota and obesity covering the period 2015-2020

Hormone or neuropeptide	Model	Aim	Intervention/procedure	Main results	Refs.
GLP-1 and PYY	HFD mice	To compare the effect of *B. animalis spp. Lactis* GLC2505 (*BlaG*) and *B. longum* JCM1217[T] (BloJ) in metabolic syndrome and obesity	C57BL/6J male mice HFD fed and supplemented with BloJ or BlaG. Faecal and blood samples collected for GLP-1, acetate and glucose tolerance determination. Microbiome assayed by 16S rRNA sequencing and qPCR.	BlaG treatment, but no BloJ, improve metabolic alteration related with obesity and exerted changes in microbioma: ↓ visceral fat accumulation ↑ plasma GLP-1, ↑ glucose tolerance ↑ genus *Bifidobacterium*	[114]
	DIO mice	To elucidate the favourable impact of probiotics *L. rhamnosus*, *L. acidophilus* and *B. bifidum* on gut microbiota, gut permeability, insulin sensitivity and peripheral and central signaling in HFD and control animals	Obesity induced-HFD mice submitted to probiotic intake (DIOPB) or not (DIOPF) fr 5 weeks. Blood extracted to glucose, insulin, LPS, cytokines and GLP-1 determination. Microbiome analysed by 16S rRNA sequencing from faeces. Liver, muscle, ileum and hypothalamus extracted to analysis of inflammatory, lipid and glucose peripheral and central.	Probiotic revert obesity-related features induced gut microbiota: ↓ intestinal permeability, ↓ LPS translocation ↓ low-grade systemic inflammation, ↑ GLP-1 release, ↓ hypothalamic insulin and leptin resistance Microbial profile: continued presence of *Bacteroidetes* ↑ *Actinobacteria*, ↓ *Firmicutes*	[115]
	Human	To examine the effect of synbiotic supplementation on metabolic syndrome.	Metabolic syndrome patient consumed for 12 weeks a weight-loss diet supplemented with:	Beneficial effects were found in the synbiotic group: ↓ BMI ↓ fasting blood sugar levels, ↑ insulin sensitivity	[169]

Table 2. (Continued)

Hormone or neuropeptide	Model	Aim	Intervention/procedure	Main results	Refs.
GLP-1 and PYY *(continued)*			1) synbiotic capsule (67 probiotic strain bacteria + fructooligosaccharide) or 2) placebo capsule. Anthropometric measurement and blood sample for PYY, GLP-1 and insulin measure taken in the beginning and end of experiment.	↑ GLP-1 ↑ PYY Continuous weight loss in symbiotic group, stopping in placebo subjects in week 6	
	HFD mice	To clarify the physiological role of barley β-glucan (BG) in obesity mouse model involving gut hormones and microbioma remodelling.	HFD obese mice divided in high BG (HBG), low BG (LBG) and control group. *Firmicutes*, *Bacteroidetes* and *Actinobacteria* abundance measured by RT-PCR in faeces, PYY and GLP-1 evaluated in plasma, and SCFAs in plasma and faeces.	Both, HBG and LBG compared to control: ↓ appetite, ↓ weight gain, ↓ fat mass, ↑ insulin sensitivity ↑ *Actinobacteria* (major acetate producer) HBG compared to LBG mice: ↑ SCFAs, ↑ PYY and ↑ GLP-1 Similar *Firmicutes:Bacteroidetes* ratio	[117]
	HFD rat	To validate inulin effects on food intake, energy expenditure, body composition, gut microbiota, and gut hormones in obesity rat model	HFD rats supplemented with 0, 10 or 25% inulin or 25% cellulose for 21 days. Food intake, energy expenditure, body composition, glucose tolerance, gut hormones and butyrate enzyme production tested.	Reduction in caloric intake is inulin dose dependent. Inulin supplementation: ↑ *Bacteroidetes* and *Bifidobacterium spp.*, ↓ *Clostridium* ↑ PYY, CCK and pro-glucagon transcript	[118]

Hormone or neuropeptide	Model	Aim	Intervention/procedure	Main results	Refs.
	Overweitgh adults (human)	To demonstrate if colonic propionate improve PYY and GLP-1 secretion in humans, and reduce energy intake and weight gain in overweight adults	Inulin-propionate ester, inulin or placebo administered for 24 weeks to overweight adults. Body weight and body composition, gut hormones (PYY, GLP-1), glucose homeostasis, lipid risk marker and gut microbiota analysed.	Inulin-propionate fermentation: ↓ food intake, ↓ body weight gain ↑ plasma PYY and GLP-1, ↑ insulin sensitivity ↓ intrahepatocellular and abdominal lipid content ↑ *Bacteroides* and *Atopobium* genus in inulin-propionate ester assay Inulin fermentation: ↑ *Bifidobacterium*	[119]
	Murine cells and mice	To decipher if H2S, produced by colonic SRB can regulate GLP-1 secretion.	Murine L-cells (GLUTag) exposed to H2S. Mice diet supplemented with prebiotic chondroitin sulphate for 4 weeks. Glucose tolerance, GLP-1 and *Desulfovibrio piger* levels estimated.	GLP-1 secretion was stimulated in culture cells. Condroitin sulphate treated animals: ↑ H2S and *Desulfovibrio piger* levels in faeces ↑ GLP-1 ↑ insulin secretion, ↑ glucose tolerance ↓ reduced food consumption	[122]
	Mice	To test the epigallocatechin-3-gallate (EGCG) effect on diet-induced obesity mice through regulation of bile acid (BA) metabolism and gut microbiota.	Pathogen-free or WT mice fed WD or control diet for 8 months and then divided receiving vehicle or EGCG during 10 months. Blood collected to insulin tolerance, GLP-1 and PYY determination. Microbial profile evaluated by 16S rRNA sequencing in stool.	EGCG shifts gut microbiota and regulates BA signalling thereby having a metabolic beneficial effect: ↓ visceral fat ↓ insulin resistance ↓ GLP-1 ↑ PYY in serum ↑ *Akkermansia muciniphila*.	[120]

Table 2. (Continued)

Hormone or neuropeptide	Model	Aim	Intervention/procedure	Main results	Refs.
GLP-1 and PYY (continued)	Rat	To study if prebiotic fibre intake in overnourished rat would promote satiety hormone and weight gain, since overnutrition during early development has been linked to metabolic disease and obesity in adulthood	Male rat coming from SL or NL were randomized in control or 10% OFS diet. Body composition, glucose tolerance, gut hormones and gut microbiota investigated.	OFS intake, independently of litter size: ↑ satiety, ↑ glucose tolerance ↑ PYY and GLP-1 secretion ↑ *Bifidobacteria*, ↓ *Akkermansia muciniphila* OFS intake in SL rats: ↓ body fat ↓ glycaemia similar to levels seen in NL rats.	[116]
	Rat	To investigate the desirable effects of Bifidobacteria (BB) and high-purity GOS (HP-GOS) supplement.	Rats divided in HP-GOS; HP-GOS+BB; BB and control. Faeces collected to bacterial composition analysis. PYY and GLP-1 gene expression measured in plasma.	Prebiotic GOS and BB exert beneficial effects. HP-GOS and HP-GOS+BB groups: ↑ Bifidobacteria HP-GOS and BB groups: ↑ PYY and ↑ GLP-1 mRNA	[121]
	HDF rat	To evaluated if berberine supplementation could improve the metabolic state of HFD rats through modulation of microbioma-gut-brain axis	Rats consumed a chow diet, HFD, and HFD + berberine groups. Blood collected for insulin, lipid profile and brain-gut peptides assay. GLP-1R expression determined in hypothalamus. Gut microbiota analysed by fecal 16S rRNA sequencing.	Berberine supplement induced: ↓ weight gain, ↓ lipolysis, ↑ orexin, ↓ NPY in plasma ↑ GLP-1, ↑ hypothalamic GLP-1R expression Microbiome profile: no differences at phyla levels. ↑ *Bacteroides*, ↑ *Aenaerofilum* ↑ *Sutterella, Bilophila, Desulfovibrio* ↓ *Blautia, Dorea and Roseburia*	[123]

Hormone or neuropeptide	Model	Aim	Intervention/procedure	Main results	Refs.
	Ob/ob mice	To validate the impact of dietary capsaicin and the alteration in gut microbiota in an obese background.	Ob/ob mice subjected to normal, low 0,01% capsaicin (LC) or high 0.02% capsaicin (HC) diet for 6 weeks. Glucose homeostasis, insulin sensitivity, SCFAs, gastrointestinal hormones, pro-inflammatory cytokines and gut microbiota composition quantified.	Capsaicin LC and HC treatment: ↑ glucose and ↑ insulin tolerance, ↑ plasma GLP-1, ↓ ghrelin, ↑ fecal butyrate ↓ TNF-α, IL-1β, and IL-6 Microbiota profile: ↑ *Firmicutes:Bacteroidetes* ratio (Phyla) ↑ *Roseburia* ↓ *Bacteroides*, ↓ *Parabacteroidetes* (genus) no significant differences between LC and HC	[124]
	DIO rat	To compare the effectiveness of OAGB and SADS with RYGB and SG as weight loss strategies in obesity.	Surgeries performed in diet-induced obese Long-Evans rats. Metabolic outcomes were monitored before and for 15 weeks after surgery. Weight loss, fat percent, glucose homeostasis, insulin and GLP-1 release, lipid metabolism (hepatic/circulating cholesterol and triglycerides) and gut microbiota alteration were measured.	All procedures improve obesity state, but: OAGB and RYGB: ↓Weight ↓fat content notably OAGB and SADS: ↑GLP-1, ↑insulin specially OAGB, SG and RUGB ↓ cholesterol; OAGB, SG ↓ circulating triglycerides OAGB and RYGB ↓ hepatic triglycerides Fecal microbiome communities significantly altered in diversity or composition, except SG.	[125]

Table 2. (Continued)

Hormone or neuropeptide	Model	Aim	Intervention/procedure	Main results	Refs.
GLP-1 and PYY (*continued*)	HFD rat	To examine the possible metabolic improvement of GG, a novel SG resecting ~80% of the glandular portion leaving the forestomach almost intact, in HFD rats.	HFD was administrated to rats for 10 weeks before and after the intervention. Profitable effects were verified at least 10 weeks after surgery. Glucose tolerance, insulin, GIT hormones, glucose and lipid metabolism, and microbial profile were stimated.	GG surgery evidenced: ↑ hepatic and peripheral insulin sensitivity ↓ ectopic and abdominal fat deposition ↓ hepatic glycogen accumulation ↑GLP-1 and ↓ ghrelin Microbioma profile: ↓*Ruminococcus*, ↑*Lactobacillus* ↑*Collinsella*	[126]
	Human (severely obese)	To elucidate if bariatric surgery promotes desirable changes in circulating gastrointestinal peptides and fecal microbiota in obese patient.	Food intake, body composition, gut hormones (PYY, GLP-1/2, orexin, ghrelin and CCK) and fecal microbioma evaluated in severely obese and normal weight control subjects at time 0 and 6 months after biliointestinal bypass.	Severely obese patient changes with surgery: ↓ BMI and ↓ weight, ↓ insulin resistance ↑ plasma GLP-1, ↑ PYY and ↑ ghrelin levels Microbial community: ↓ *Clostridial XIVa* and *IV* clusters ↑ *Lactobacillus crispatus*, ↑ *Meghasphaera elsdenii*	[127]

Hormone or neuropeptide	Model	Aim	Intervention/procedure	Main results	Refs.
CCK	HFD mice	To determine the beneficial effects of 2´-fucosyllactose (2´FL) supplementation on microbiota-gut-brain axis and obese features.	Mice consumed low fat diet, HF diet, and HF diet + 1,2,5 and 10% 2´FL for 6 weeks. Body weight, energy intake, fat and lean mass, gut-brain signalling, and faecal microbiome and metabolites were assayed	Only 10% 2´FL group: ↓ fat mass, ↓ body weight gain ↑ *Parabacteroides* Food intake inhibition by CCK lost in HF-diet was restored.	[128]
	HFD rat	To study the possible role of potato resistant starch (RS) supplementation in microbioma composition, inflammatory status and vagal signalling in HF diet rats.	Rats divided in chow diet, high fat (HF) and high fat supplemented with RS (HFRS). Food intake, body weight, inflammation markers, glucose tolerance, CCK sensitivity, vagal afferent activation and microbioma profile were research.	Microbiota manipulation via RS enrichment diet: ↓ hyperphagia ↓ weight gain ↑ CCK-induced satiety restored glucose homeostasis restore obesity-related vagal phenotype ↓ prevent inflammatory status Restored microbiota composition, more closely related to chow than HF diet: ↑ *Bifidobacterium spp.*	[129]
	DIO rats	To clarify if HPD reverse neuroinflammation, CCK sensitivity, glucose levels and gut microbiome from an obese background	Rats fed first with WD diet (high fat and sugar) for 12 weeks and later with HPD for 6 weeks. Body composition, food intake, CCK sensitivity, inflammation, glucose levels, brain signalling and microbiota were determined.	HPD diet, after obese-induced diet, produced: ↓ body fat, ↑ CCK sensitivity restore glucose homeostasis ↑ *Akkermansia muciniphila* in correlation with fat mass loss no change in brain inflammation were noticed.	[130]

Table 2. (Continued)

Hormone or neuropeptide	Model	Aim	Intervention/procedure	Main results	Refs.
CCK *(continued)*	HFD rat, human	To investigate if thykaloids affect gastrointestinal passage and microbial composition in HFD rats and human.	HFD rats and human volunteers subjected to placebo or thykaloid intake. Gastric emptying and intestinal transit studied in rat. Gut microbiome composition analysed in human.	Dietary supplementation with thykaloids affects satiety via gastrointestinal fullness and modulating microbiota composition suggesting its role to prevent and treat obesity. ↓ gastric emptying ↓ intestinal transit ↑ *Bacteroides fragilis*	[131]
Ghrelin	DIO rats	Explore the effect of resistant starch (RS) or exogenous acetate carried by RS (RSA) in the gut on metabolic syndrome throught GMPA and GMPB production.	RS or exogenous acetate carrier by RS (RSA) administered to diet-induced obese rats. Faecal metabolomics SCFAs analysed, lipid metabolism genes, microbiota profile and gut hormones were examined.	RSA treatment, compared to RS: ↓ body weight gain, ↓ lipogenesis gene expression ↓ ghrelin secretion, ↑ insulin sensitivity ↑ butyrate ↑ GMPB: *Coprococcus, Faecalibacterium, Roseburia* and *Eubacterium*	[132]
	DIO rats	To understand the effects of cinnamaldehyde (bioactive cinnamon component) consumption in HFD obese-related altered glucose and lipid metabolism, microbial composition alteration	Cinnamaldehyde was dispensed to diet-induced obese rats. Fasting hyperphagia, gut hormones plasma levels, adipose tissue lipolysis, inflammation marker and cecal microbial composition stimated.	Cinnamaldehyde prevent HFD-induced ↓ body weight gain, ↓ fasting hyperphagia normalized circulating levels of leptin/ghrelin ratio: ↑ ghrelin and ↓ leptin ↑ adipose tissue lipolysis, ↓ IL-1β level	[133]

Hormone or neuropeptide	Model	Aim	Intervention/procedure	Main results	Refs.
		and gut hormone regulation.		No differences we observed in selected *Lactobacillus, Bifidobacteria,* and *Roseburia* gut microbial analysis	
	HDF rat	To elucidate if changes in acetate producer gut microbiota is associated with insulin resistance and metabolic syndrome leading to obesity.	Rats divided in HFD or control groups. FMT performed between HFD and control. SCFAs levels and gut hormones were assayed.	Altered gut microbiota promote obesity in rodent: ↑ acetate production, ↑ ghrelin release ↑ glucose-stimulated insulin secretion Control subject submitted to faecal transplantation from HFD: ↑ higher acetate and ↑ insulin levels ↑ *Firmicutes:Bacteroidetes* ratio	[170]
Leptin	Newborn rats	To determine the influence of leptin supplementation on the intraepithelial lymphocyte composition, gut barrier, gene expression and gut microbiota profile	Daily leptin supplementation to newborn Wistar rats during the first 21 days of life. Lymphocyte composition, intestinal gene expression and cecal microbiota checked.	Leptin supplementation in suckling period had an impact on both intraepithelial lymphocyte and gut microbiota pattern: ↑ CD8αα⁺ intraepithelial lymphocytes, ↑ TNF-α, ↓ FcR ↑ MUC-2 and ↑ MUC-3 expression, ↓ *Proteobacteria* phylum and *Blautia* ↓ *Sutterella*, ↑ *Clostridium* genus	[146]
	Mice	To investigate the role of leptin signaling in microbiota composition and its impact in the stomach pathogenesis	C57 BL/6 WT mice, leptin receptor (Lepr)-mutated db/db mice and gastrointestinal epithelium-specific leptin receptor KO mice (T3 b-*Lepr* KO) were fed with HFD or control diet.	Gastric dysbiosis in HFD mice increased gastric leptin leading to metaplasia. Host microbiota transplantation from: HFD WT mice: induce metaplasia in host mice	[145]

Table 2. (Continued)

Hormone or neuropeptide	Model	Aim	Intervention/procedure	Main results	Refs.
Leptin (continued)			T3 b-*Lepr* KO mice were administered with recombinant leptin. Gastric microbiota from three groups were transplanted to recipient mice.	T3 b-*Lepr* KO HFD mice: presented limited damage in the host stomach and failed in develop spontaneous obesity db/db and T3 b-*Lepr* KO HFD: did not present changes in host gastric microbiota or intestinal metaplasia.[	
	Human	To evaluate and compare the bacterial richness and diversity on fecal microbiota	Mexican school-age undernourished, obese and normal weight children gut microbiome were analysed by 16s rRNA sequencing.	Positive correlation of *Lachnospiraceae* with leptin and negative with energy consumption Positive correlation of *Proteobacteria* with total fat intake	[171]
	Human	To decipher relations between leptin, insulin and the taxonomic and functional potentials of the infant microbiome. To identify early differences in gut microbiota between infants born from obese or normal weight mothers	Breast feeding infants born from obese or normal-weight mothers. Stool collected to SCFAs assay and gut microbiota 16s rRNA sequencing and whole genome sequencing. Insulin and leptin quantified in human milk.	Human milk insulin and leptin were independently associated with microbial pathways involved in the increase of intestinal barrier function and reduction of intestinal inflammation	[149]

Hormone or neuropeptide	Model	Aim	Intervention/procedure	Main results	Refs.
	ob/ob mice	To evaluate the effect of inulin supplementation on the cecal microbiota	Inulin-supplemented diet in C57BL/6J leptin-deficient ob/ob mice. Cecal transcriptome and microbiome diversity composition were stimated.	Inulin intake in ob/ob mice: ↓ α-diversity; ↑ β-diversity (similar to WT) ↑ *Prevotellaceae* Improvement of glucose and lipid metabolism Reversion in changes associated with leptin signaling pathways, especially AMPK signaling mediated by gut microbiota.	[141]
	HFD ob/ob mice	To validate if estrogens and leptin modulate gut microbiota	HFD leptin-deficient ob/ob (ovariectomized) female mice with E2 implants for 35 days and leptin supplementation	E2 and leptin are responsible of changes in gut microbiota ↑ abundance of S24-7 family by E2 ↑ *coriobacteriaceae, Clostridium* and *Lactobacillus* by leptin	[172]
	DIO rats	To clarify the effects of resveratrol in fat accumulation and its relation with leptin sensitivity To identify which resveratrol-derived circulating metabolites are potentially involved in its effect	Daily consumption of resveratrol (50 mg/kg or 100 mg/kg or 200 mg/kg) in DIO Wister rats.Body weight, fat content, resveratrol metabolites, leptin plasma levels and leptin signaling pathways were evaluated.	Higher doses of resveratrol: ↓ body weight ↓ fat accumulation. ↑ leptin periphery sensitivity Higher resveratrol metabolites found are derived from microbiota metabolism	[142]

Table 2. (Continued)

Hormone or neuropeptide	Model	Aim	Intervention/procedure	Main results	Refs.
Leptin *(continued)*	HFD mice	To test the effects of the lack of gut microbiota on body weight, and in the expression and the methylation of the leptin promoter	Exogenous leptin administration for 7 days to HFD C57 BL/6 J conventional and GF mice.	Increased expression of fat hydrolysis and oxidative genes in conventional mice but not in those lacking microbiota	[144]
	HFD/HSD obese mice	To explore the relationship among diet supplements, gut microbiota, host genetics and metabolic status.	C57 BL/6 J leptin-deficient ob/ob mice fed with standard diet, HFD or high sucrose diet	Disturbs in the structure of gut microbiota particularly elevate under leptin deficient conditions.	[147]
	HFD ob/ob obese mice	To examine the role of the gut microbiota and leptin as mediators in the anhedonia state due to HFD	Selective gut bacteria depletion in C57 BL/6 J leptin-deficient ob/ob HFD mice with non-absorbable antibiotics. Behavioral tests and molecular analysis (gut microbiome composition, intestinal metabolome, fecal fatty acids, plasma hormone levels) performed.	Anhedonia state was associated with gut microbiome involving leptin as a signaling. Antibiotic HFD mice: ↓ leptin, ↓ weight gain, ↑ self care ↓ SCFAs and gut microbe related metabolites	[173]
	HFD Mice	To investigate the kind of dietary fat that causes intestinal metaplasia	C57 BL/6 mice fed with HDF containing: lard, beef tallow, hydrogenated coconut oil, linseed oil, corn oil, olive oil, soybean oil	Especially lard and coconut: ↑ leptin and ↑ TNF- α level Lard and linseed-type: ↑ tumorigenicity	[137]

Hormone or neuropeptide	Model	Aim	Intervention/procedure	Main results	Refs.
			or cocoa butter and water ad libitum for three months. Carcinogen 1-methyl-3-nitro-1-nitrosoguanidine were also administered twice a week	Microbiome dysbiosis was not related with obesity or the onset of intestinal metaplasia Pathogenesis of intestinal metaplasia seen to be depended on gastric leptin signaling mediated through beta-catenin signaling	
	HFD mice	To evaluate the role of *Bifidobacterium pseudocatenulatum* CECT 7765 in adverse endocrine and neurobehavioral consequences of obesity.	WT C57BL-6 mice fed a standard diet or HFD, supplemented with either placebo or bifidobacterial strain for 13 weeks. Behavioural tests, corticosterone, glucocorticoid receptor in hypothalamous, hippocampus and intestine, TLR2, monoamines, insulin and leptin evaluated.	*B. pseudocatenulatum* treatment on obese mice: → anhedonia, ↓depressive-like behaviour → exaggerated HPA stress response ↓ hyperleptinemia, ↓ corticosterone levels ↑ glucocorticoids and leptin receptors in hippocampus	[139]
	Ob/ob mice	To determine the association between non-HFD-induced obesity and dysbiosis and their impact in gut permeability	C57 BL/6 J leptin-deficient ob/ob. Comparative and correlative analyses of gut microbiome composition, gut permeability, intestinal structural changes, tight junction-mucin formation and cellular turnover tested.	Ob/ob mice, compared to control: ↑ Intestinal turnover, ↑ Gut permeability ↓ tight junctions: ↓ occludin, ↓ Zo1, ↓ Jam-forming genes ↓ mucus sinthesys: ↓ Muc2 and Muc6 microbiome dysbiosis: ↑ *Firmicutes:Bacteroidetes*	[174]
	Over-nourishing human infants	To identify the role of the infant gut microbiota in infants with excessive	Breast fed infants with excessive weight gain. Faeces and milk samples collected to 16s rRNA sequencing.	Excessive weight gain during breastfeeding: ↓ Enterococus,	[148]

Table 2. (Continued)

Hormone or neuropeptide	Model	Aim	Intervention/procedure	Main results	Refs.
Leptin (continued)		weight gain during breast feeding	Plasma leptin quantified.	↑ BMI, ↑ waist circumference ↑ body fat, ↑ leptin Inverse correlation between infant plasma levels and *Enterococcus* abundance, both in infant and breast milk	
	Obese human adults	To compare the metabolome of obese patients after RYGB surgery	Euglycemic or diabetics obese human adults submitted to RYGB. Metabolomic profile were analysed.	RYGB patients after surgery: ↓ Leptin immediately, ↓ weight loss reveal strong association of leptin with leucine, tryptophan, complex fatty acids and lipids no association of leptin with carbohydrates	[143]
Omentin-1	Obese human children	To explore the effect of *B. pseudocatenulatum* CECT 7765 on cardiometabolic risk factors, inflammatory cytokines and gut microbiota composition	Dietary counseling and *B. pseudocatenulatum* CECT 7765 administrated at a single daily dose of 1×109–10 CFU for 13 weeks in obese children with insulin resistance	↓ Circulating hs-CRP and MCP-1 ↑ HDL-C and omentin-1 Changes in microbiota were minors and correlations between modified bacterial groups and omentin-1 were not detected	[142]
Insulin	Obese humans	To demonstrate the effect of antibiotics treatment on fasting and postprandial forearm muscle substrate metabolism and	Amoxicillin (broad-spectrum antibiotic) or vancomycin (narrow-spectrum antibiotic) were served to obese human with insulin resistance for 7 days (1,500 mg/day)	Antibiotics treatment no affected: fasting or postprandial plasma glucose, triacylglycerol, glycerol, lactate and insulin Vancomycin:	[163]

Hormone or neuropeptide	Model	Aim	Intervention/procedure	Main results	Refs.
		postprandial insulin sensitivity		↓ bacterial diversity, ↓ Gram-positive *Firmicutes* ↑ Gram-negative *Proteobacteria* abundance Amoxicillin: did not affect microbial composition	
	Obese and T2D humans	To study the fecal microbiota profile in T2D patients and to evaluate changes in composition after surgery.	Randomized T2D and obese patients were submitted to Roux-en-Y duodeno jejunostomy with a minimal proximal vertical gastrectomy	Surgery improves microbial richness and abundance, especially in bacteria associated with weight loss and metabolism control bacteria	[162]
	HFD mice	To examine the impact of the medicinal mushroom Antrodia cinnamomea in obesity	C57BL/6J wild type or HFD mice were treated with *A. cinnamomea* in water for 8 weeks	↓ body weight ↓ insulin resistance ↓ inflammation	[156]
	Human Overweight and obese adults	To check associations between induced improvement in insulin sensitivity and changes in host biology, gut microbiota and lifestyle factors.	Overweight or obese humans adults were calorie restricted for 6 weeks. Lifestyle factors (diet and physical activity), sAT gene expression, metabolomics of serum, urine and feces, and gut microbiota composition evaluated.	Highlight insulin sensitivity associated with: serum branched chain amino acids, sAT stress and ubiquitination-related genes changes in gut microbiota profile	[159]
	Mice	To identify alterations in both the composition and diurnal oscillation of gut microbiota in obesity after SG.	C57 BL/6 J WT consumed HFD or chow diet for 14 weeks and submitted to SG or sham surgery. Faecal samples collected for 16s rRNA sequencing.	After sleeve gastrectomy surgery: ↑ weight loss, ↑ glucose tolerance, ↑ insulin sensitivity restauration of richness and composition as well as the diurnal oscillation of gut microbiota	[161]

Table 2. (Continued)

Hormone or neuropeptide	Model	Aim	Intervention/procedure	Main results	Refs.
Insulin *(continued)*	Overweight/ obese human	To clarify the association between circulating ceramides with alteration in gut microbiome.	Overweight and obese with diabetes risk eat low calorie diet with soluble fibers, high protein intake and low glycemic index carbohydrates for 6 weeks. Serum extracted to HPLC-MS/MS sphingolipid assay and stool for shotgun metagenomic sequencing.	Serum cermides correlated with: ↑ fasting glucose, ↓ insulin sensitivity ↓ β-cell function. ↓ gut microbiota richness: ↑ lipopolysaccharide synthesis genes ↓ methanogenesis and bile acid metabolism genes	[175]
	Obese human adults	To assess the effects of walnuts on cardiometabolic outcomes	Smoothie containing 48 g walnuts administered to obese adults twice, with 1 month washout period between them	Walnut consumption exhibit longer benefit: ↓ glucose, ↓ insulin, ↓ insulin resistance, ↑ PYY no changes in cardiometabolic markers	[158]
	Human: Obese children	To elucidate mechanisms implied in the predisposition toward hyperinsulinemia and its complications in obese children	Pre-pubertal obese children with insulin resistance or not. Untargeted metabolomic LC-MS, GC-MS and CE-MS analysis from serum.	Insulin resistance was associated to alterations in inflammation and central carbon metabolism: ↑ bile acids, ↑ branched chain amino acids and metabolites In presence of hyperinsulinemia: ↑ pyruvate, ↓ ketone, ↑ glycolysis, ↓ 3-hydroxybutyrate	[176]

Hormone or neuropeptide	Model	Aim	Intervention/procedure	Main results	Refs.
	Humans: Obese adults	To test the effect and mechanism of daily intake of *L. reuteri* in insulin sensitivity, cytokine release and insulin secretion.	*L. reuteri* SD5865 administered for 4 weeks to glucose-tolerant and obese adults. Glucose tolerance test, incretin hormones, insulin, cytokines and lipid content tested	No changes in alpha diversity between obese and leans control No changes in peripheral and hepatic insulin sensitivity, body mass, ectopic fat content, or circulating cytokines	[155]
	HFD mice	To determine the effect of several types of dietary fiber and different SCFA ratios in metabolic syndrome	B6 HFD mice supplemented with: 10% dietary fiber o SCFA: 10% cellulose (HFC); 3% cellulose + 7% inulin (HFI); 3% cellulose + 7% guar gum (HFG); 5% cellulose + Acetate:Propionate 10:1 (HAc); and 5% cellulose + Acetate:Propionate 1:2.5 (HPr) during 30 weeks	Inulin and SCFA: Improve Insulin resistance ↓ HFD-induced body weight/fat gain ↑ *B. animalis*, particularly stimulated by inulin ↑ *B. pseudolongum*, particularly stimulated by guar gum	[157]
GH	Mice and bovine	To examine the influence of GH on the mouse gut microbiome	Gut microbiota analysis in GH gene disrupted mice (GH –/-) with C57BL/6J background and transgenic bovines with chronic excessive GH	GH altered the gut microbial composition GH showed minimal effect on microbial diversity yet significantly altered maturity	[166]

Table 2. (Continued)

Hormone or neuropeptide	Model	Aim	Intervention/procedure	Main results	Refs.
GC	Obese human	To investigate the association between long-term GC therapy in obesity development and related impaired microbiota.	16S rRNA sequencing and gas chromatography of SCFAs determined in faecal content of healthy and glucocorticoid-treated obese individuals.	GC-treated obese people: ↓ SCFAs ↓ microbiota diversity, ↓*Bacteroidetes* ↑ *Firmicutes* (specially genus *Streptococcus*)	[168]

↑ increase or enhancing in; ↓ reduction or diminution in;

16s rRNA: 16S ribosomal ribonucleic acid; 2'FL: 2´-fucosyllactose; BA: bile acids; BB: bifidobacteria; BMI: body mass index; CCK: cholecystokinin; CE-MS: capillary electrophoresis-mass spectrometry; CFU: colony forming units; DIO: diet-induced obesity; E2: 17β-estradiol; EGCG: epigallocatechin-3-gallate; FcR: fragment cristalizable region receptor; FMT: fecal microbiome transplantation; GC-MS: gas chromatography-mass spectrometry; GC: glucocorticoid; GF: germ free; GH: growth hormone; GG: glandular gastrectomy; GLP-1: glucagon-like peptide 1; GLP-1R: glucagon-like peptide 1 receptor; GMPA: gut microbiota-produced acetate; GMPB: gut microbiota-produced butyrate; GOS: galactooligosaccharides; H2S: hydrogen sulphide; HDL-C: high-density lipoprotein cholesterol; HFD: high fat diet; HPA: hypothalamic-pituitary axis; HPD: high protein diet; HPLC-MS/MS: high-performance liquid chromatography-tandem mass spectrometry; hs-CRP: high-sensitive C-reactive protein; IL-1β: interleukin 1 beta; IL-6: interleukin 6; KO: knockout; LC-MS: liquid chromatography- mass spectrometry; LPS: lipopolysaccharides; NL: normal litter; MCP-1: monocyte chemoattractant protein-1; OAGB: one-anastomosis gastric bypass; OFS: oligofructose; PYY: polypeptide YY; qPCR: quantitative PCR; RT-PCR: real-time PCR; RS: resistant starch; RYGB: Roux-en-Y gastric bypass; SADS: single-anastomosis duodenal switch; sAT: subcutaneous adipose tissue, SCFAs: short-chain fatty acids; SG: sleeve gastrectomy; SL: small litter; SRB: sulphate-reducing bacteria; T2D: type 2 diabetes; TLR2: toll-like receptor 2; TNF-α: tumor necrosis factor-alpha; WD: westernized diet; WT: wild type.

CONCLUSION

Obesity is closely related to genetic factors but also to lifestyle, diet and the imbalance between food intake and the amount of energy consumed by the individual [177]. Evidence shows that the gut-brain axis and alterations in the gut microbiota are deeply related to obesity and have an important role in bidirectional interactions between the gut and the nervous system. Since obesity and associated diseases have become a public health problem, studies addressed to highlight new insights into the role of gut microbiota in its pathogenesis and mechanisms as well as therapeutic perspectives have increased during the last years. It is accepted that different dietary patterns and feeding behaviors exert an important impact on gut microbiota profile [178] but, the mechanism about how gut microbiota affects neuroendocrine function is not deeply understood. Medical advances (i.e., bariatric surgery or fecal microbiota transplantation), lifestyle changes and other interventions that also target the gut microbiota (i.e., probiotics, antibiotics, nutritional supplements, etc.,) are being investigated because of their effectiveness in treating obesity. However, such strategies do not provide a global overview to shed light on complex triad compounds by gut microbiota, obesity and neuroendocrine system. In consequence, there is a deep need for understanding the ability of the gut microbiota to alter the release of both brain and gut hormones and/or related-peptides and the impact in the homeostasis and pathways implied in the development of obesity. From a medical care point of view, the main goal could be to restore gut microbiota by the modulation of pathways controlled by the neuroendocrine system and identify those signaling with a direct role in regulating appetite. In this regard, strategies targeted pathways related to the gut-hypothalamic axis and anorexic or orexigenic hormones, looking for altering the composition of the microbiota to achieve a specific microbial profile or the other way around could be effective in the management of obesity and its related complications. Also, it is worthy to remark the need to investigate associations between obese state, microbiota and inflammation, since this is a potential mechanism by which bacteria connect with the brain, including bacterial sub-products that gain access to the brain by the bloodstream, via cytokine release from mucosal immune cells, via the release of the gut.

To conclude, the knowledge about the link between the gut-brain-microbiota axis and its role in both promoting or regulating energy and its contribution to the obesogenic state is still scarce. Unfortunately, this hinders the design of interventions in which the composition of the microbiota may be modulated to achieve a specific beneficial microbial profile. In consequence more research in this regard is mandatory.

REFERENCES

[1]   Health Topics, *Obesity*. n.d. https://www.who.int/topics/obesity/en/.

[2]   Ghoorah, K., Campbell, P., Kent, A., Maznyczka, A. and Kunadian, V. (2016). Obesity and cardiovascular outcomes: a review. *European heart journal. Acute cardiovascular care*, 5(1): 77–85. https://doi.org/ 10.1177/2048872614523349.

[3]   Cussotto, S., Sandhu, K. V., Dinan, T. G. and Cryan, J. F. (2018). The Neuroendocrinology of the Microbiota-Gut-Brain Axis: A Behavioural Perspective. *Frontiers in neuroendocrinology*, 51: 80–101. https://doi.org/10.1016/j.yfrne.2018.04.002.

[4]   Wu, H. J. and Wu, E. (2012). The role of gut microbiota in immune homeostasis and autoimmunity. *Gut microbes*, 3(1): 4–14. https://doi. org/10.4161/gmic.19320.

[5]   Mayer E. A. (2011). Gut feelings: the emerging biology of gut-brain communication. *Nature reviews. Neuroscience*, 12(8): 453–466. https://doi.org/10.1038/nrn3071.

[6]   Weltens, N., Iven, J., Van Oudenhove, L. and Kano, M. (2018). The gut-brain axis in health neuroscience: implications for functional gastrointestinal disorders and appetite regulation. *Annals of the New York Academy of Sciences*, 1428(1): 129–150. https://doi.org/10. 1111/nyas.13969.

[7]   Gore, A. C. (2013). Neuroendocrine System. In: Squire, L., Berg, D., Bloom, F. E., du Lac, S., Ghosh, A., Spitzer, N. C. (eds.).

Fundamental Neuroscience (4th Edition). Academic Press. 799-817. https://doi.org/10.1016/B978-0-12-385870-2.00038-X.

[8] Nishiyama, Y. and Katsura, K. (2015). The Neuroendocrine System and Its Regulation. In: Uchino, H., Ushijima, K., Ikeda Y. (eds.). *Neuroanesthesia and Cerebrospinal Protection.* Springer, Tokyo. 31-38. https://doi.org/10.1007/978-4-431-54490-6_3.

[9] Ahlman, H. and Nilsson (2001). The gut as the largest endocrine organ in the body. *Annals of oncology: official journal of the European Society for Medical Oncology,* 12 Suppl 2, S63–S68. https://doi.org/10.1093/annonc/12.suppl_2.s63.

[10] Berthoud, H. R. and Morrison, C. (2008). The brain, appetite, and obesity. *Annual review of psychology,* 59: 55–92. https://doi.org/10.1146/annurev.psych.59.103006.093551.

[11] Gropp, E., Shanabrough, M., Borok, E., Xu, A. W., Janoschek, R., Buch, T., Plum, L., Balthasar, N., Hampel, B., Waisman, A., Barsh, G. S., Horvath, T. L. and Brüning, J. C. (2005). Agouti-related peptide-expressing neurons are mandatory for feeding. *Nature neuroscience,* 8(10): 1289–1291. https://doi.org/10.1038/nn1548.

[12] Morton, G. J., Cummings, D. E., Baskin, D. G., Barsh, G. S. and Schwartz, M. W. (2006). Central nervous system control of food intake and body weight. *Nature,* 443(7109): 289–295. https://doi.org/10.1038/nature05026.

[13] Rodríguez, E. M., Blázquez, J. L. and Guerra, M. (2010). The design of barriers in the hypothalamus allows the median eminence and the arcuate nucleus to enjoy private milieus: the former opens to the portal blood and the latter to the cerebrospinal fluid. *Peptides,* 31(4): 757–776. https://doi.org/10.1016/j.peptides.2010.01.003.

[14] Suzuki, K., Jayasena, C. N., and Bloom, S. R. (2012). Obesity and appetite control. *Experimental diabetes research,* 824305. https://doi.org/10.1155/2012/824305.

[15] Volkow, N. D., Wang, G. J. and Baler, R. D. (2011). Reward, dopamine and the control of food intake: implications for obesity. *Trends in cognitive sciences,* 15(1): 37–46. https://doi.org/10.1016/j.tics.2010.11.001.

[16] Furness, J. B., Callaghan, B. P., Rivera, L. R. and Cho, H. J. (2014). The enteric nervous system and gastrointestinal innervation: integrated local and central control. *Advances in experimental medicine and biology*, 817: 39–71. https://doi.org/10.1007/978-1-4939-0897-4_3.

[17] Crowley, V. E., Yeo, G. S. and O'Rahilly, S. (2002). Obesity therapy: altering the energy intake-and-expenditure balance sheet. *Nature reviews. Drug discovery*, 1(4): 276–286. https://doi.org/10.1038/nrd770.

[18] Calzada-León, R., Altamirano-Bustamante, N. and Ruiz-Reyes M. L. (2018). Reguladores Neuroendocrinos y Gastrointestinales Del Apetito y La Saciedad. *Boletín Medico del Hospital Infantil de México*, 65(6): 468-487. [Neuroendocrine and Gastrointestinal Regulators of Appetite and Satiety. *Medical Bulletin of the Children's Hospital of Mexico*, 65(6): 468-487].

[19] Timper, K. and Brüning, J. C. (2017). Hypothalamic circuits regulating appetite and energy homeostasis: pathways to obesity. *Disease models & mechanisms*, 10(6): 679–689. https://doi.org/10.1242/dmm.026609.

[20] Sobrino Crespo, C., Perianes Cachero, A., Puebla Jiménez, L., Barrios, V. and Arilla Ferreiro, E. (2014). Peptides and food intake. *Frontiers in endocrinology*, 5, 58. https://doi.org/10.3389/fendo.2014.00058.

[21] Steinert, R. E., Feinle-Bisset, C., Asarian, L., Horowitz, M., Beglinger, C. and Geary, N. (2017). Ghrelin, CCK, GLP-1, and PYY(3-36): Secretory Controls and Physiological Roles in Eating and Glycemia in Health, Obesity, and After RYGB. *Physiological reviews*, 97(1): 411–463. https://doi.org/10.1152/physrev.00031.2014.

[22] Lean, M. E. and Malkova, D. (2016). Altered gut and adipose tissue hormones in overweight and obese individuals: cause or consequence?. *International journal of obesity*, 40(4): 622–632. https://doi.org/10.1038/ijo.2015.220.

[23] Willms, B., Werner, J., Holst, J. J., Orskov, C., Creutzfeldt, W. and Nauck, M. A. (1996). Gastric emptying, glucose responses, and insulin secretion after a liquid test meal: effects of exogenous glucagon-like peptide-1 (GLP-1)-(7-36) amide in type 2 (noninsulin-dependent) diabetic patients. *The Journal of clinical endocrinology and metabolism*, 81(1): 327–332. https://doi.org/10.1210/jcem.81.1. 8550773.

[24] Turton, M. D., O'Shea, D., Gunn, I., Beak, S. A., Edwards, C. M., Meeran, K., Choi, S. J., Taylor, G. M., Heath, M. M., Lambert, P. D., Wilding, J. P., Smith, D. M., Ghatei, M. A., Herbert, J. and Bloom, S. R. (1996). A role for glucagon-like peptide-1 in the central regulation of feeding. *Nature*, 379(6560): 69–72. https://doi.org/10.1038/ 379069a0.

[25] Alhadeff, A. L., Rupprecht, L. E. and Hayes, M. R. (2012). GLP-1 neurons in the nucleus of the solitary tract project directly to the ventral tegmental area and nucleus accumbens to control for food intake. *Endocrinology*, 153(2): 647–658. https://doi.org/10.1210/en. 2011-1443.

[26] Müller, T. D., Finan, B., Bloom, S. R., D'Alessio, D., Drucker, D. J., Flatt, P. R., Fritsche, A., Gribble, F., Grill, H. J., Habener, J. F., Holst, J. J., Langhans, W., Meier, J. J., Nauck, M. A., Perez-Tilve, D., Pocai, A., Reimann, F., Sandoval, D. A., Schwartz, T. W., Seeley, R. J., Stemmer, K., Tang-Christensen, M., Woods, S. C., DiMarchi, R. D. and Tschoep, M. H. (2019). Glucagon-like peptide 1 (GLP-1). *Molecular metabolism*, 30: 72–130. https://doi.org/10.1016/j.molmet. 2019.09.010.

[27] Ranganath, L. R., Beety, J. M., Morgan, L. M., Wright, J. W., Howland, R. and Marks, V. (1996). Attenuated GLP-1 secretion in obesity: cause or consequence?. *Gut*, 38(6): 916–919. https://doi.org/ 10.1136/gut.38.6.916.

[28] Adam, T. C. and Westerterp-Plantenga, M. S. (2005). Glucagon-like peptide-1 release and satiety after a nutrient challenge in normal-weight and obese subjects. *The British journal of nutrition*, 93(6): 845–851. https://doi.org/10.1079/bjn20041335.

[29] Verdich, C., Toubro, S., Buemann, B., Lysgård Madsen, J., Juul Holst, J. and Astrup, A. (2001). The role of postprandial releases of insulin and incretin hormones in meal-induced satiety--effect of obesity and weight reduction. *International journal of obesity and related metabolic disorders: journal of the International Association for the Study of Obesity*, 25(8): 1206–1214. https://doi.org/10.1038/sj.ijo.0801655.

[30] Li, P., Tiwari, H. K., Lin, W. Y., Allison, D. B., Chung, W. K., Leibel, R. L., Yi, N. and Liu, N. (2014). Genetic association analysis of 30 genes related to obesity in a European American population. *International journal of obesity* (2005), 38(5): 724–729. https://doi.org/10.1038/ijo.2013.140.

[31] Verdich, C., Flint, A., Gutzwiller, J. P., Näslund, E., Beglinger, C., Hellström, P. M., Long, S. J., Morgan, L. M., Holst, J. J. and Astrup, A. (2001). A meta-analysis of the effect of glucagon-like peptide-1 (7-36) amide on ad libitum energy intake in humans. *The Journal of clinical endocrinology and metabolism*, 86(9): 4382–4389. https://doi.org/10.1210/jcem.86.9.7877.

[32] Batterham, R. L., Cowley, M. A., Small, C. J., Herzog, H., Cohen, M. A., Dakin, C. L., Wren, A. M., Brynes, A. E., Low, M. J., Ghatei, M. A., Cone, R. D. and Bloom, S. R. (2002). Gut hormone PYY (3-36) physiologically inhibits food intake. *Nature*, 418(6898): 650–654. https://doi.org/10.1038/nature00887.

[33] Koda, S., Date, Y., Murakami, N., Shimbara, T., Hanada, T., Toshinai, K., Niijima, A., Furuya, M., Inomata, N., Osuye, K. and Nakazato, M. (2005). The role of the vagal nerve in peripheral PYY3-36-induced feeding reduction in rats. *Endocrinology*, 146(5): 2369–2375. https://doi.org/10.1210/en.2004-1266.

[34] Holzer, P., Reichmann, F. and Farzi, A. (2012). Neuropeptide Y, peptide YY and pancreatic polypeptide in the gut-brain axis. *Neuropeptides,* 46(6), 261–274. https://doi.org/10.1016/j.npep.2012.08.005.

[35] Zwirska-Korczala, K., Konturek, S. J., Sodowski, M., Wylezol, M., Kuka, D., Sowa, P., Adamczyk-Sowa, M., Kukla, M., Berdowska, A.,

Rehfeld, J. F., Bielanski, W. and Brzozowski, T. (2007). Basal and postprandial plasma levels of PYY, ghrelin, cholecystokinin, gastrin and insulin in women with moderate and morbid obesity and metabolic syndrome. *Journal of physiology and pharmacology: an official journal of the Polish Physiological Society*, 58 Suppl 1, 13–35.

[36] Batterham, R. L., Cohen, M. A., Ellis, S. M., Le Roux, C. W., Withers, D. J., Frost, G. S., Ghatei, M. A. and Bloom, S. R. (2003). Inhibition of food intake in obese subjects by peptide YY3-36. *The New England journal of medicine*, 349(10): 941–948. https://doi.org/10.1056/NEJMoa030204.

[37] Sloth, B., Holst, J. J., Flint, A., Gregersen, N. T. and Astrup, A. (2007). Effects of PYY1-36 and PYY3-36 on appetite, energy intake, energy expenditure, glucose and fat metabolism in obese and lean subjects. *American journal of physiology. Endocrinology and metabolism*, 292(4): E1062–E1068. https://doi.org/10.1152/ajpendo.00450.2006.

[38] Fried, M., Erlacher, U., Schwizer, W., Löchner, C., Koerfer, J., Beglinger, C., Jansen, J. B., Lamers, C. B., Harder, F. and Bischof-Delaloye, A. (1991). Role of cholecystokinin in the regulation of gastric emptying and pancreatic enzyme secretion in humans. Studies with the cholecystokinin-receptor antagonist loxiglumide. *Gastroenterology*, 101(2): 503–511. https://doi.org/10.1016/0016-5085(91)90031-f.

[39] Schwartz, G. J. and Moran, T. H. (1994). CCK elicits and modulates vagal afferent activity arising from gastric and duodenal sites. *Annals of the New York Academy of Sciences*, 713: 121–128. https://doi.org/10.1111/j.1749-6632.1994.tb44058.x.

[40] Beglinger, C., Degen, L., Matzinger, D., D'Amato, M. and Drewe, J. (2001). Loxiglumide, a CCK-A receptor antagonist, stimulates calorie intake and hunger feelings in humans. *American journal of physiology. Regulatory, integrative and comparative physiology*, 280(4): R1149–R1154. https://doi.org/10.1152/ajpregu.2001.280.4.R1149.

[41] Rehfeld J. F. (2017). Cholecystokinin-From Local Gut Hormone to Ubiquitous Messenger. *Frontiers in endocrinology*, 8, 47. https://doi.org/10.3389/fendo.2017.00047.

[42] Marchal-Victorion, S., Vionnet, N., Escrieut, C., Dematos, F., Dina, C., Dufresne, M., Vaysse, N., Pradayrol, L., Froguel, P. and Fourmy, D. (2002). Genetic, pharmacological and functional analysis of cholecystokinin-1 and cholecystokinin-2 receptor polymorphism in type 2 diabetes and obese patients. *Pharmacogenetics*, 12(1): 23–30. https://doi.org/10.1097/00008571-200201000-00004.

[43] de Lartigue, G., Barbier de la Serre, C., Espero, E., Lee, J. and Raybould, H. E. (2012). Leptin resistance in vagal afferent neurons inhibits cholecystokinin signaling and satiation in diet induced obese rats. *PloS one*, 7(3), e32967. https://doi.org/10.1371/journal.pone.0032967.

[44] Dodd, G. T. and Tiganis, T. (2017). Insulin action in the brain: Roles in energy and glucose homeostasis. *Journal of neuroendocrinology*, 29(10), 10.1111/jne.12513. https://doi.org/10.1111/jne.12513.

[45] Cowley, M. A., Smart, J. L., Rubinstein, M., Cerdán, M. G., Diano, S., Horvath, T. L., Cone, R. D. and Low, M. J. (2001). Leptin activates anorexigenic POMC neurons through a neural network in the arcuate nucleus. *Nature*, 411(6836): 480–484. https://doi.org/10.1038/35078085.

[46] Rezai-Zadeh, K., Yu, S., Jiang, Y., Laque, A., Schwartzenburg, C., Morrison, C. D., Derbenev, A. V., Zsombok, A. and Münzberg, H. (2014). Leptin receptor neurons in the dorsomedial hypothalamus are key regulators of energy expenditure and body weight, but not food intake. *Molecular metabolism*, 3(7): 681–693. https://doi.org/10.1016/j.molmet.2014.07.008.

[47] Scacchi, M., Pincelli, A. I. and Cavagnini, F. (1999). Growth hormone in obesity. *International journal of obesity and related metabolic disorders: journal of the International Association for the Study of Obesity*, 23(3): 260–271. https://doi.org/10.1038/sj.ijo.0800807.

[48] Pan, H., Guo, J. and Su, Z. (2014). Advances in understanding the interrelations between leptin resistance and obesity. *Physiology & behavior*, 130: 157–169. https://doi.org/10.1016/j.physbeh.2014.04. 003.

[49] Enriori, P. J., Evans, A. E., Sinnayah, P., Jobst, E. E., Tonelli-Lemos, L., Billes, S. K., Glavas, M. M., Grayson, B. E., Perello, M., Nillni, E. A., Grove, K. L. and Cowley, M. A. (2007). Diet-induced obesity causes severe but reversible leptin resistance in arcuate melanocortin neurons. *Cell metabolism*, 5(3): 181–194. https://doi.org/10.1016/j. cmet.2007.02.004.

[50] Carlson, J. J., Turpin, A. A., Wiebke, G., Hunt, S. C. and Adams, T. D. (2009). Pre- and post- prandial appetite hormone levels in normal weight and severely obese women. *Nutrition & metabolism*, 6, 32. https://doi.org/10.1186/1743-7075-6-32.

[51] Ozata, M., Ozdemir, I. C. and Licinio, J. (1999). Human leptin deficiency caused by a missense mutation: multiple endocrine defects, decreased sympathetic tone, and immune system dysfunction indicate new targets for leptin action, greater central than peripheral resistance to the effects of leptin, and spontaneous correction of leptin-mediated defects. *The Journal of clinical endocrinology and metabolism*, 84(10): 3686–3695. https://doi.org/10.1210/jcem.84.10. 5999.

[52] Yang, R. Z., Lee, M. J., Hu, H., Pray, J., Wu, H. B., Hansen, B. C., Shuldiner, A. R., Fried, S. K., McLenithan, J. C. and Gong, D. W. (2006). Identification of omentin as a novel depot-specific adipokine in human adipose tissue: possible role in modulating insulin action. *American journal of physiology. Endocrinology and metabolism*, 290(6): E1253–E1261. https://doi.org/10.1152/ajpendo.00572.2004.

[53] Watanabe, T., Watanabe-Kominato, K., Takahashi, Y., Kojima, M. and Watanabe, R. (2017). Adipose Tissue-Derived Omentin-1 Function and Regulation. *Comprehensive Physiology*, 7(3): 765–781. https://doi.org/10.1002/cphy.c160043.

[54] Herder, C., Ouwens, D. M., Carstensen, M., Kowall, B., Huth, C., Meisinger, C., Rathmann, W., Roden, M. and Thorand, B. (2015).

Adiponectin may mediate the association between omentin, circulating lipids and insulin sensitivity: results from the KORA F4 study. *European journal of endocrinology*, 172(4): 423–432. https://doi.org/10.1530/EJE-14-0879.

[55]   de Souza Batista, C. M., Yang, R. Z., Lee, M. J., Glynn, N. M., Yu, D. Z., Pray, J., Ndubuizu, K., Patil, S., Schwartz, A., Kligman, M., Fried, S. K., Gong, D. W., Shuldiner, A. R., Pollin, T. I. and McLenithan, J. C. (2007). Omentin plasma levels and gene expression are decreased in obesity. *Diabetes*, 56(6): 1655–1661. https://doi.org/10.2337/db06-1506.

[56]   Zhang, Q., Zhu, L., Zheng, M., Fan, C., Li, Y., Zhang, D., He, Y. and Yang, H. (2014). Changes of serum omentin-1 levels in normal subjects, type 2 diabetes and type 2 diabetes with overweight and obesity in Chinese adults. *Annales d'endocrinologie*, 75(3): 171–175. https://doi.org/10.1016/j.ando.2014.04.013.

[57]   Fukuhara, A., Matsuda, M., Nishizawa, M., Segawa, K., Tanaka, M., Kishimoto, K., Matsuki, Y., Murakami, M., Ichisaka, T., Murakami, H., Watanabe, E., Takagi, T., Akiyoshi, M., Ohtsubo, T., Kihara, S., Yamashita, S., Makishima, M., Funahashi, T., Yamanaka, S., Hiramatsu, R., … Shimomura, I. (2005). Visfatin: a protein secreted by visceral fat that mimics the effects of insulin. *Science*, 307(5708): 426–430. https://doi.org/10.1126/science.1097243.

[58]   Sonoli, S. S., Shivprasad, S., Prasad, C. V., Patil, A. B., Desai, P. B. and Somannavar, M. S. (2011). Visfatin--a review. *European review for medical and pharmacological sciences*, 15(1): 9–14.

[59]   Curat, C. A., Wegner, V., Sengenès, C., Miranville, A., Tonus, C., Busse, R. and Bouloumié, A. (2006). Macrophages in human visceral adipose tissue: increased accumulation in obesity and a source of resistin and visfatin. *Diabetologia*, 49(4): 744–747. https://doi.org/10.1007/s00125-006-0173-z.

[60]   Stastny, J., Bienertova-Vasku, J. and Vasku, A. (2012). Visfatin and its role in obesity development. *Diabetes & metabolic syndrome*, 6(2): 120–124. https://doi.org/10.1016/j.dsx.2012.08.011.

[61] Kumari, B. and Yadav, U. (2018). Adipokine Visfatin's Role in Pathogenesis of Diabesity and Related Metabolic Derangements. *Current molecular medicine*, 18(2), 116–125. https://doi.org/10.2174/1566524018666180705114131.

[62] Zhang, J. V., Ren, P. G., Avsian-Kretchmer, O., Luo, C. W., Rauch, R., Klein, C. and Hsueh, A. J. (2005). Obestatin, a peptide encoded by the ghrelin gene, opposes ghrelin's effects on food intake. *Science,* 310(5750): 996–999. https://doi.org/10.1126/science.1117255.

[63] Lacquaniti, A., Donato, V., Chirico, V., Buemi, A., abd Buemi, M. (2011). Obestatin: an interesting but controversial gut hormone. *Annals of nutrition & metabolism*, 59(2-4): 193–199. https://doi.org/ 10.1159/000334106.

[64] Trovato, L., Gallo, D., Settanni, F., Gesmundo, I., Ghigo, E. and Granata, R. (2014). Obestatin: is it really doing something?. *Frontiers of hormone research*, 42: 175–185. https://doi.org/10.1159/ 000358346.

[65] Huda, M. S., Durham, B. H., Wong, S. P., Deepak, D., Kerrigan, D., McCulloch, P., Ranganath, L., Pinkney, J. and Wilding, J. P. (2008). Plasma obestatin levels are lower in obese and post-gastrectomy subjects, but do not change in response to a meal. *International journal of obesity*, 32(1), 129–135. https://doi.org/10.1038/sj.ijo. 0803694.

[66] Zamrazilová, H., Hainer, V., Sedláčková, D., Papezová, H., Kunesová, M., Bellisle, F., Hill, M. and Nedvídková, J. (2008). Plasma obestatin levels in normal weight, obese and anorectic women. *Physiological research*, 57 Suppl 1, S49–S55.

[67] Nakahara, T., Harada, T., Yasuhara, D., Shimada, N., Amitani, H., Sakoguchi, T., Kamiji, M. M., Asakawa, A. and Inui, A. (2008). Plasma obestatin concentrations are negatively correlated with body mass index, insulin resistance index, and plasma leptin concentrations in obesity and anorexia nervosa. *Biological psychiatry*, 64(3): 252–255. https://doi.org/10.1016/j.biopsych.2007.08.005.

[68] Keire, D. A., Kobayashi, M., Solomon, T. E. and Reeve, J. R. (2000). Solution structure of monomeric peptide YY supports the functional

significance of the PP-fold. *Biochemistry*, 39(32): 9935–9942. https://doi.org/10.1021/bi992576a.

[69] Batterham, R. L., Le Roux, C. W., Cohen, M. A., Park, A. J., Ellis, S. M., Patterson, M., Frost, G. S., Ghatei, M. A. and Bloom, S. R. (2003). Pancreatic polypeptide reduces appetite and food intake in humans. *The Journal of clinical endocrinology and metabolism*, 88(8): 3989–3992. https://doi.org/10.1210/jc.2003-030630.

[70] Lin, S., Shi, Y. C., Yulyaningsih, E., Aljanova, A., Zhang, L., Macia, L., Nguyen, A. D., Lin, E. J., During, M. J., Herzog, H. and Sainsbury, A. (2009). Critical role of arcuate Y4 receptors and the melanocortin system in pancreatic polypeptide-induced reduction in food intake in mice. *PloS one*, 4(12), e8488. https://doi.org/10.1371/journal.pone.0008488.

[71] Asakawa, A., Inui, A., Yuzuriha, H., Ueno, N., Katsuura, G., Fujimiya, M., Fujino, M. A., Niijima, A., Meguid, M. M. and Kasuga, M. (2003). Characterization of the effects of pancreatic polypeptide in the regulation of energy balance. *Gastroenterology*, 124(5): 1325–1336. https://doi.org/10.1016/s0016-5085(03)00216-6.

[72] Lassmann, V., Vague, P., Vialettes, B. and Simon, M. C. (1980). Low plasma levels of pancreatic polypeptide in obesity. *Diabetes*, 29(6): 428–430. https://doi.org/10.2337/diab.29.6.428.

[73] Reinehr, T., Enriori, P. J., Harz, K., Cowley, M. A. and Roth, C. L. (2006). Pancreatic polypeptide in obese children before and after weight loss. *International journal of obesity*, 30(10): 1476–1481.

[74] Könner, A. C., Janoschek, R., Plum, L., Jordan, S. D., Rother, E., Ma, X., Xu, C., Enriori, P., Hampel, B., Barsh, G. S., Kahn, C. R., Cowley, M. A., Ashcroft, F. M. and Brüning, J. C. (2007). Insulin action in AgRP-expressing neurons is required for suppression of hepatic glucose production. *Cell metabolism*, 5(6): 438–449. https://doi.org/10.1016/j.cmet.2007.05.004.

[75] Dodd, G. T., Decherf, S., Loh, K., Simonds, S. E., Wiede, F., Balland, E., Merry, T. L., Münzberg, H., Zhang, Z. Y., Kahn, B. B., Neel, B. G., Bence, K. K., Andrews, Z. B., Cowley, M. A. and Tiganis, T. (2015). Leptin and insulin act on POMC neurons to promote the

browning of white fat. *Cell*, 160(1-2): 88–104. https://doi.org/10.1016/j.cell.2014.12.022.

[76] Heni, M., Kullmann, S., Preissl, H., Fritsche, A. and Häring, H. U. (2015). Impaired insulin action in the human brain: causes and metabolic consequences. *Nature reviews. Endocrinology*, 11(12): 701–711. https://doi.org/10.1038/nrendo.2015.173.

[77] Weaver, J. U., Kopelman, P. G. and Hitman, G. A. (1992). Central obesity and hyperinsulinaemia in women are associated with polymorphism in the 5' flanking region of the human insulin gene. *European journal of clinical investigation*, 22(4): 265–270. https://doi.org/10.1111/j.1365-2362.1992.tb01461.x.

[78] Shin, A. C., Filatova, N., Lindtner, C., Chi, T., Degann, S., Oberlin, D. and Buettner, C. (2017). Insulin Receptor Signaling in POMC, but Not AgRP, Neurons Controls Adipose Tissue Insulin Action. *Diabetes*, 66(6): 1560–1571. https://doi.org/10.2337/db16-1238.

[79] Lutz, T. A., Mollet, A., Rushing, P. A., Riediger, T. and Scharrer, E. (2001). The anorectic effect of a chronic peripheral infusion of amylin is abolished in area postrema/nucleus of the solitary tract (AP/NTS) lesioned rats. *International journal of obesity and related metabolic disorders: journal of the International Association for the Study of Obesity*, 25(7): 1005–1011. https://doi.org/10.1038/sj.ijo.0801664.

[80] Rushing, P. A., Hagan, M. M., Seeley, R. J., Lutz, T. A. and Woods, S. C. (2000). Amylin: a novel action in the brain to reduce body weight. *Endocrinology*, 141(2): 850–853. https://doi.org/10.1210/endo.141.2.7378.

[81] Reda, T. K., Geliebter, A. and Pi-Sunyer, F. X. (2002). Amylin, food intake, and obesity. *Obesity research*, 10(10): 1087–1091. https://doi.org/10.1038/oby.2002.147.

[82] Reinehr, T., de Sousa, G., Niklowitz, P. and Roth, C. L. (2007). Amylin and its relation to insulin and lipids in obese children before and after weight loss. *Obesity*, 15(8): 2006–2011. https://doi.org/10.1038/oby.2007.239.

[83] Geary, N., Kissileff, H. R., Pi-Sunyer, F. X. and Hinton, V. (1992). Individual, but not simultaneous, glucagon and cholecystokinin

infusions inhibit feeding in men. *The American journal of physiology*, 262(6 Pt 2): R975–R980. https://doi.org/10.1152/ajpregu.1992.262. 6.R975.

[84] Heppner, K. M., Habegger, K. M., Day, J., Pfluger, P. T., Perez-Tilve, D., Ward, B., Gelfanov, V., Woods, S. C., DiMarchi, R. and Tschöp, M. (2010). Glucagon regulation of energy metabolism. *Physiology & behavior*, 100(5): 545–548. https://doi.org/10.1016/j.physbeh.2010. 03.019.

[85] Day, J. W., Ottaway, N., Patterson, J. T., Gelfanov, V., Smiley, D., Gidda, J., Findeisen, H., Bruemmer, D., Drucker, D. J., Chaudhary, N., Holland, J., Hembree, J., Abplanalp, W., Grant, E., Ruehl, J., Wilson, H., Kirchner, H., Lockie, S. H., Hofmann, S., Woods, S. C., Nogueiras, R., Pfluger, P. T., Perez-Tilve, D., DiMarchi, R. and Tschöp, M. H. (2009). A new glucagon and GLP-1 co-agonist eliminates obesity in rodents. *Nature chemical biology*, 5(10): 749–757. https://doi.org/10.1038/nchembio.209.

[86] Al-Massadi, O., Fernø, J., Diéguez, C., Nogueiras, R. and Quiñones, M. (2019). Glucagon Control on Food Intake and Energy Balance. *International journal of molecular sciences*, 20(16), 3905. https://doi. org/10.3390/ijms20163905.

[87] Stern, J. H., Smith, G. I., Chen, S., Unger, R. H., Klein, S. and Scherer, P. E. (2019). Obesity dysregulates fasting-induced changes in glucagon secretion. *The Journal of endocrinology*, 243(2): 149–160. https://doi.org/10.1530/JOE-19-0201.

[88] Manell, H., Kristinsson, H., Kullberg, J., Ubhayasekera, S., Mörwald, K., Staaf, J., Cadamuro, J., Zsoldos, F., Göpel, S., Sargsyan, E., Ahlström, H., Bergquist, J., Weghuber, D., Forslund, A. and Bergsten, P. (2019). Hyperglucagonemia in youth is associated with high plasma free fatty acids, visceral adiposity, and impaired glucose tolerance. *Pediatric diabetes*, 20(7): 880–891. https://doi.org/10. 1111/pedi.12890.

[89] Korbonits, M., Goldstone, A. P., Gueorguiev, M. and Grossman, A. B. (2004). Ghrelin--a hormone with multiple functions. *Frontiers in*

neuroendocrinology, 25(1): 27–68. https://doi.org/10.1016/j.yfrne. 2004.03.002.

[90] Nakazato, M., Murakami, N., Date, Y., Kojima, M., Matsuo, H., Kangawa, K. and Matsukura, S. (2001). A role for ghrelin in the central regulation of feeding. *Nature*, 409(6817): 194–198. https://doi.org/10.1038/35051587.

[91] Chen, H. Y., Trumbauer, M. E., Chen, A. S., Weingarth, D. T., Adams, J. R., Frazier, E. G., Shen, Z., Marsh, D. J., Feighner, S. D., Guan, X. M., Ye, Z., Nargund, R. P., Smith, R. G., Van der Ploeg, L. H., Howard, A. D., MacNeil, D. J. and Qian, S. (2004). Orexigenic action of peripheral ghrelin is mediated by neuropeptide Y and agouti-related protein. *Endocrinology*, 145(6): 2607–2612. https://doi.org/10.1210/en.2003-1596.

[92] Broglio, F., Arvat, E., Benso, A., Gottero, C., Muccioli, G., Papotti, M., van der Lely, A. J., Deghenghi, R. and Ghigo, E. (2001). Ghrelin, a natural GH secretagogue produced by the stomach, induces hyperglycemia and reduces insulin secretion in humans. *The Journal of clinical endocrinology and metabolism*, 86(10): 5083–5086. https://doi.org/10.1210/jcem.86.10.8098.

[93] Briggs, D. I., Enriori, P. J., Lemus, M. B., Cowley, M. A. and Andrews, Z. B. (2010). Diet-induced obesity causes ghrelin resistance in arcuate NPY/AgRP neurons. *Endocrinology*, 151(10): 4745–4755. https://doi.org/10.1210/en.2010-0556.

[94] Burdyga, G., Varro, A., Dimaline, R., Thompson, D. G. and Dockray, G. J. (2006). Ghrelin receptors in rat and human nodose ganglia: putative role in regulating CB-1 and MCH receptor abundance. *American journal of physiology. Gastrointestinal and liver physiology*, 290(6): G1289–G1297. https://doi.org/10.1152/ajpgi.00543.2005.

[95] Osterstock, G., Mitutsova, V., Barre, A., Granier, M., Fontanaud, P., Chazalon, M., Carmignac, D., Robinson, I. C., Low, M. J., Plesnila, N., Hodson, D. J., Mollard, P. and Méry, P. F. (2016). Somatostatin triggers rhythmic electrical firing in hypothalamic GHRH neurons. *Scientific reports*, 6, 24394. https://doi.org/10.1038/srep24394.

[96] Vijayakumar, A., Yakar, S. and Leroith, D. (2011). The intricate role of growth hormone in metabolism. *Frontiers in endocrinology*, 2, 32. https://doi.org/10.3389/fendo.2011.00032.

[97] Ahmad, I., Finkelstein, J. A., Downs, T. R. and Frohman, L. A. (1993). Obesity-associated decrease in growth hormone-releasing hormone gene expression: a mechanism for reduced growth hormone mRNA levels in genetically obese Zucker rats. *Neuroendocrinology*, 58(3): 332–337. https://doi.org/10.1159/000126558.

[98] Stanley, T. L. and Grinspoon, S. K. (2015). Effects of growth hormone-releasing hormone on visceral fat, metabolic, and cardiovascular indices in human studies. *Growth hormone & IGF research: official journal of the Growth Hormone Research Society and the International IGF Research Society*, 25(2): 59–65. https://doi.org/10.1016/j.ghir.2014.12.005.

[99] Brazeau, P., Vale, W., Burgus, R., Ling, N., Butcher, M., Rivier, J. and Guillemin, R. (1973). Hypothalamic polypeptide that inhibits the secretion of immunoreactive pituitary growth hormone. *Science (New York, N.Y.)*, 179(4068): 77–79. https://doi.org/10.1126/science.179. 4068.77.

[100] Stengel, A. and Taché, Y. (2019). Central somatostatin signaling and regulation of food intake. *Annals of the New York Academy of Sciences*, 1455(1): 98–104. https://doi.org/10.1111/nyas.14178.

[101] Brunani, A., Invitti, C., Dubini, A., Piccoletti, R., Bendinelli, P., Maroni, P., Pezzoli, G., Ramella, G., Calogero, A. and Cavagnini, F. (1995). Cerebrospinal fluid and plasma concentrations of SRIH, beta-endorphin, CRH, NPY and GHRH in obese and normal weight subjects. *International journal of obesity and related metabolic disorders: journal of the International Association for the Study of Obesity*, 19(1): 17–21.

[102] Kumar, U. and Singh, S. (2020). Role of Somatostatin in the Regulation of Central and Peripheral Factors of Satiety and Obesity. *International journal of molecular sciences*, 21(7), 2568. https://doi.org/10.3390/ijms21072568.

[103] Steyn, F. J., Tolle, V., Chen, C. and Epelbaum, J. (2016). Neuroendocrine Regulation of Growth Hormone Secretion. *Comprehensive Physiology*, 6(2): 687–735. https://doi.org/10.1002/cphy.c150002.

[104] Bates, P. C., Loughna, P. T., Pell, J. M., Schulster, D. and Millward, D. J. (1993). Interactions between growth hormone and nutrition in hypophysectomized rats: body composition and production of insulin-like growth factor-I. *The Journal of endocrinology*, 139(1): 117–126. https://doi.org/10.1677/joe.0.1390117.

[105] Erman, A., Veilleux, A., Tchernof, A. and Goodyer, C. G. (2011). Human growth hormone receptor (GHR) expression in obesity: I. GHR mRNA expression in omental and subcutaneous adipose tissues of obese women. *International journal of obesity,* 35(12), 1511–1519. https://doi.org/10.1038/ijo.2011.23.

[106] Pasarica, M., Zachwieja, J. J., Dejonge, L., Redman, S. and Smith, S. R. (2007). Effect of growth hormone on body composition and visceral adiposity in middle-aged men with visceral obesity. *The Journal of clinical endocrinology and metabolism*, 92(11): 4265–4270. https://doi.org/10.1210/jc.2007-0786.

[107] Uchoa, E. T., Aguilera, G., Herman, J. P., Fiedler, J. L., Deak, T. and de Sousa, M. B. (2014). Novel aspects of glucocorticoid actions. *Journal of neuroendocrinology*, 26(9): 557–572. https://doi.org/10.1111/jne.12157.

[108] Akalestou, E., Genser, L. and Rutter, G. A. (2020). Glucocorticoid Metabolism in Obesity and Following Weight Loss. *Frontiers in endocrinology*, 11, 59. https://doi.org/10.3389/fendo.2020.00059.

[109] Livingstone, D. E., Jones, G. C., Smith, K., Jamieson, P. M., Andrew, R., Kenyon, C. J. and Walker, B. R. (2000). Understanding the role of glucocorticoids in obesity: tissue-specific alterations of corticosterone metabolism in obese Zucker rats. *Endocrinology*, 141(2): 560–563. https://doi.org/10.1210/endo.141.2.7297.

[110] Swali, A., Walker, E. A., Lavery, G. G., Tomlinson, J. W. and Stewart, P. M. (2008). 11beta-Hydroxysteroid dehydrogenase type 1 regulates insulin and glucagon secretion in pancreatic islets.

Diabetologia, 51(11): 2003–2011. https://doi.org/10.1007/s00125-008-1137-2.

[111] Zakrzewska, K. E., Cusin, I., Stricker-Krongrad, A., Boss, O., Ricquier, D., Jeanrenaud, B. and Rohner-Jeanrenaud, F. (1999). Induction of obesity and hyperleptinemia by central glucocorticoid infusion in the rat. *Diabetes*, 48(2): 365–370. https://doi.org/10.2337/diabetes.48.2.365.

[112] Agustí, A., García-Pardo, M. P., López-Almela, I., Campillo, I., Maes, M., Romaní-Pérez, M. and Sanz, Y. (2018). Interplay Between the Gut-Brain Axis, Obesity and Cognitive Function. *Frontiers in neuroscience*, 12, 155. https://doi.org/10.3389/fnins.2018.00155.

[113] Pucci, A. and Batterham, R. L. (2019). Mechanisms underlying the weight loss effects of RYGB and SG: similar, yet different. *Journal of endocrinological investigation*, 42(2): 117–128. https://doi.org/10.1007/s40618-018-0892-2.

[114] Aoki, R., Kamikado, K., Suda, W., Takii, H., Mikami, Y., Suganuma, N., Hattori, M. and Koga, Y. (2017). A proliferative probiotic Bifidobacterium strain in the gut ameliorates progression of metabolic disorders via microbiota modulation and acetate elevation. *Scientific reports*, 7, 43522. https://doi.org/10.1038/srep43522.

[115] Bagarolli, R. A., Tobar, N., Oliveira, A. G., Araújo, T. G., Carvalho, B. M., Rocha, G. Z., Vecina, J. F., Calisto, K., Guadagnini, D., Prada, P. O., Santos, A., Saad, S. and Saad, M. (2017). Probiotics modulate gut microbiota and improve insulin sensitivity in DIO mice. *The Journal of nutritional biochemistry*, 50: 16–25. https://doi.org/10.1016/j.jnutbio.2017.08.006.

[116] Reid, D. T., Eller, L. K., Nettleton, J. E. and Reimer, R. A. (2016). Postnatal prebiotic fibre intake mitigates some detrimental metabolic outcomes of early overnutrition in rats. *European journal of nutrition*, 55(8): 2399–2409. https://doi.org/10.1007/s00394-015-1047-2.

[117] Miyamoto, J., Watanabe, K., Taira, S., Kasubuchi, M., Li, X., Irie, J., Itoh, H. and Kimura, I. (2018). Barley β-glucan improves metabolic condition via short-chain fatty acids produced by gut microbial

fermentation in high fat diet fed mice. *PloS one*, 13(4), e0196579. https://doi.org/10.1371/journal.pone.0196579.

[118] Singh, A., Zapata, R. C., Pezeshki, A., Reidelberger, R. D. and Chelikani, P. K. (2018). Inulin fiber dose-dependently modulates energy balance, glucose tolerance, gut microbiota, hormones and diet preference in high-fat-fed male rats. *The Journal of nutritional biochemistry*, 59: 142–152. https://doi.org/10.1016/j.jnutbio.2018.05. 017.

[119] Chambers, E. S., Viardot, A., Psichas, A., Morrison, D. J., Murphy, K. G., Zac-Varghese, S. E., MacDougall, K., Preston, T., Tedford, C., Finlayson, G. S., Blundell, J. E., Bell, J. D., Thomas, E. L., Mt-Isa, S., Ashby, D., Gibson, G. R., Kolida, S., Dhillo, W. S., Bloom, S. R., Morley, W. and Frost, G. (2015). Effects of targeted delivery of propionate to the human colon on appetite regulation, body weight maintenance and adiposity in overweight adults. *Gut*, 64(11): 1744–1754. https://doi.org/10.1136/gutjnl-2014-307913.

[120] Sheng, L., Jena, P. K., Liu, H. X., Hu, Y., Nagar, N., Bronner, D. N., Settles, M. L., Bäumler, A. J. and Wan, Y. Y. (2018). Obesity treatment by epigallocatechin-3-gallate-regulated bile acid signaling and its enriched Akkermansia muciniphila. *FASEB journal: official publication of the Federation of American Societies for Experimental Biology*, 32(12), fj201800370R. https://doi.org/10.1096/fj. 201800370R.

[121] Hong, K. B., Kim, J. H., Kwon, H. K., Han, S. H., Park, Y. and Suh, H. J. (2016). Evaluation of Prebiotic Effects of High-Purity Galactooligosaccharides *in vitro* and *in vivo*. *Food technology and biotechnology*, 54(2): 156–163. https://doi.org/10.17113/ftb.54.02. 16.4292.

[122] Pichette, J., Fynn-Sackey, N. and Gagnon, J. (2017). Hydrogen Sulfide and Sulfate Prebiotic Stimulates the Secretion of GLP-1 and Improves Glycemia in Male Mice. *Endocrinology*, 158(10): 3416–3425. https://doi.org/10.1210/en.2017-00391.

[123] Sun, H., Wang, N., Cang, Z., Zhu, C., Zhao, L., Nie, X., Cheng, J., Xia, F., Zhai, H. and Lu, Y. (2016). Modulation of Microbiota-Gut-

Brain Axis by Berberine Resulting in Improved Metabolic Status in High-Fat Diet-Fed Rats. *Obesity facts*, 9(6): 365–378. https://doi.org/10.1159/000449507.

[124] Song, J. X., Ren, H., Gao, Y. F., Lee, C. Y., Li, S. F., Zhang, F., Li, L. and Chen, H. (2017). Dietary Capsaicin Improves Glucose Homeostasis and Alters the Gut Microbiota in Obese Diabetic ob/ob Mice. *Frontiers in physiology*, 8, 602. https://doi.org/10.3389/fphys.2017.00602.

[125] Arble, D. M., Evers, S. S., Bozadjieva, N., Frikke-Schmidt, H., Myronovych, A., Lewis, A., Toure, M. H. and Seeley, R. J. (2018). Metabolic comparison of one-anastomosis gastric bypass, single-anastomosis duodenal-switch, Roux-en-Y gastric bypass, and vertical sleeve gastrectomy in rat. *Surgery for obesity and related diseases: official journal of the American Society for Bariatric Surgery*, 14(12): 1857–1867. https://doi.org/10.1016/j.soard.2018.08.019.

[126] Basso, N., Soricelli, E., Castagneto-Gissey, L., Casella, G., Albanese, D., Fava, F., Donati, C., Tuohy, K., Angelini, G., La Neve, F., Severino, A., Kamvissi-Lorenz, V., Birkenfeld, A. L., Bornstein, S., Manco, M. and Mingrone, G. (2016). Insulin Resistance, Microbiota, and Fat Distribution Changes by a New Model of Vertical Sleeve Gastrectomy in Obese Rats. *Diabetes*, 65(10): 2990–3001. https://doi.org/10.2337/db16-0039.

[127] Federico, A., Dallio, M., Tolone, S., Gravina, A. G., Patrone, V., Romano, M., Tuccillo, C., Mozzillo, A. L., Amoroso, V., Misso, G., Morelli, L., Docimo, L. and Loguercio, C. (2016). Gastrointestinal Hormones, Intestinal Microbiota and Metabolic Homeostasis in Obese Patients: Effect of Bariatric Surgery. *In vivo,* 30(3): 321–330.

[128] Lee, S., Goodson, M., Vang, W., Kalanetra, K., Barile, D. and Raybould, H. (2020). 2'-fucosyllactose Supplementation Improves Gut-Brain Signaling and Diet-Induced Obese Phenotype and Changes the Gut Microbiota in High Fat-Fed Mice. *Nutrients*, 12(4), E1003. https://doi.org/10.3390/nu12041003.

[129] Klingbeil, E. A., Cawthon, C., Kirkland, R. and de La Serre, C. B. (2019). Potato-Resistant Starch Supplementation Improves

Microbiota Dysbiosis, Inflammation, and Gut-Brain Signaling in High Fat-Fed Rats. *Nutrients*, 11(11), 2710. https://doi.org/10.3390/nu11112710.

[130] Wang, L., Jacobs, J. P., Lagishetty, V., Yuan, P. Q., Wu, S. V., Million, M., Reeve, J. R., Jr, Pisegna, J. R. and Taché, Y. (2017). High-protein diet improves sensitivity to cholecystokinin and shifts the cecal microbiome without altering brain inflammation in diet-induced obesity in rats. *American journal of physiology. Regulatory, integrative and comparative physiology*, 313(4): R473–R486. https://doi.org/10.1152/ajpregu.00105.2017.

[131] Stenblom, E. L., Weström, B., Linninge, C., Bonn, P., Farrell, M., Rehfeld, J. F. and Montelius, C. (2016). Dietary green-plant thylakoids decrease gastric emptying and gut transit, promote changes in the gut microbial flora, but does not cause steatorrhea. *Nutrition & metabolism*, 13, 67. https://doi.org/10.1186/s12986-016-0128-4.

[132] Si, X., Shang, W., Zhou, Z., Strappe, P., Wang, B., Bird, A. and Blanchard, C. (2018). Gut Microbiome-Induced Shift of Acetate to Butyrate Positively Manages Dysbiosis in High Fat Diet. *Molecular nutrition & food research*, 62(3), 10.1002/mnfr.201700670. https://doi.org/10.1002/mnfr.201700670.

[133] Khare, P., Jagtap, S., Jain, Y., Baboota, R. K., Mangal, P., Boparai, R. K., Bhutani, K. K., Sharma, S. S., Premkumar, L. S., Kondepudi, K. K., Chopra, K. and Bishnoi, M. (2016). Cinnamaldehyde supplementation prevents fasting-induced hyperphagia, lipid accumulation, and inflammation in high-fat diet-fed mice. *BioFactors*, 42(2): 201–211. https://doi.org/10.1002/biof.1265.

[134] Álvarez-Mercado, A. I., Bujaldon, E., Gracia-Sancho, J. and Peralta, C. (2018). The Role of Adipokines in Surgical Procedures Requiring Both Liver Regeneration and Vascular Occlusion. *International journal of molecular sciences*, 19(11): 3395. https://doi.org/10.3390/ijms19113395.

[135] Miranda, V., Dos Santos Amorim, P. R., Bastos, R. R., de Faria, E. R., de Castro Moreira, M. E., do Carmo Castro Franceschini, S., do Carmo Gouveia Peluzio, M., de Luces Fortes Ferreira, C. L. and

Priore, S. E. (2019). Abundance of Gut Microbiota, Concentration of Short-Chain Fatty Acids, and Inflammatory Markers Associated with Elevated Body Fat, Overweight, and Obesity in Female Adolescents. *Mediators of inflammation*, 7346863. https://doi.org/10.1155/2019/7346863.

[136] Tanaka, M. and Itoh, H. (2019). Hypertension as a Metabolic Disorder and the Novel Role of the Gut. *Current hypertension reports*, 21(8): 63. https://doi.org/10.1007/s11906-019-0964-5.

[137] Arita, S., Ogawa, T., Murakami, Y., Kinoshita, Y., Okazaki, M. and Inagaki-Ohara, K. (2019). Dietary Fat-Accelerating Leptin Signaling Promotes Protumorigenic Gastric Environment in Mice. *Nutrients*, 11(9): 2127. https://doi.org/10.3390/nu11092127.

[138] Varga, O., Harangi, M., Olsson, I. A. and Hansen, A. K. (2010). Contribution of animal models to the understanding of the metabolic syndrome: a systematic overview. *Obesity reviews: an official journal of the International Association for the Study of Obesity*, 11(11): 792–807. https://doi.org/10.1111/j.1467-789X.2009.00667.x.

[139] Agusti, A., Moya-Pérez, A., Campillo, I., Montserrat-de la Paz, S., Cerrudo, V., Perez-Villalba, A. and Sanz, Y. (2018). Bifidobacterium pseudocatenulatum CECT 7765 Ameliorates Neuroendocrine Alterations Associated with an Exaggerated Stress Response and Anhedonia in Obese Mice. *Molecular neurobiology*, 55(6): 5337–5352. https://doi.org/10.1007/s12035-017-0768-z.

[140] Zhao, L., Zhang, Q., Ma, W., Tian, F., Shen, H. and Zhou, M. (2017). A combination of quercetin and resveratrol reduces obesity in high-fat diet-fed rats by modulation of gut microbiota. *Food & function*, 8(12): 4644–4656. https://doi.org/10.1039/c7fo01383c.

[141] Song, X., Zhong, L., Lyu, N., Liu, F., Li, B., Hao, Y., Xue, Y., Li, J., Feng, Y., Ma, Y., Hu, Y. and Zhu, B. (2019). Inulin Can Alleviate Metabolism Disorders in ob/ob Mice by Partially Restoring Leptin-related Pathways Mediated by Gut Microbiota. *Genomics, proteomics & bioinformatics*, 17(1): 64–75. https://doi.org/10.1016/j.gpb.2019.03.001.

[142] Ardid-Ruiz, A., Ibars, M., Mena, P., Del Rio, D., Muguerza, B., Bladé, C., Arola, L., Aragonès, G. and Suárez, M. (2018). Potential Involvement of Peripheral Leptin/STAT3 Signaling in the Effects of Resveratrol and Its Metabolites on Reducing Body Fat Accumulation. *Nutrients*, 10(11), 1757. https://doi.org/10.3390/nu10111757.

[143] Seridi, L., Leo, G. C., Dohm, G. L., Pories, W. J. and Lenhard, J. (2018). Time course metabolome of Roux-en-Y gastric bypass confirms correlation between leptin, body weight and the microbiome. *PloS one*, 13(5), e0198156. https://doi.org/10.1371/journal.pone.0198156.

[144] Yao, H., Fan, C., Fan, X., Lu, Y., Wang, Y., Wang, R., Tang, T. and Qi, K. (2020). Effects of gut microbiota on leptin expression and body weight are lessened by high-fat diet in mice. *The British journal of nutrition*, 1–11. https://doi.org/10.1017/S0007114520001117.

[145] Arita, S. and Inagaki-Ohara, K. (2019). High-fat-diet-induced modulations of leptin signaling and gastric microbiota drive precancerous lesions in the stomach. *Nutrition*, 67-68, 110556. https://doi.org/10.1016/j.nut.2019.110556.

[146] Grases-Pintó, B., Abril-Gil, M., Castell, M., Rodríguez-Lagunas, M. J., Burleigh, S., Fåk Hållenius, F., Prykhodko, O., Pérez-Cano, F. J. and Franch, À. (2019). Influence of Leptin and Adiponectin Supplementation on Intraepithelial Lymphocyte and Microbiota Composition in Suckling Rats. *Frontiers in immunology*, 10, 2369. https://doi.org/10.3389/fimmu.2019.02369.

[147] Wang, J. H., Shin, N. R., Lim, S. K., Im, U., Song, E. J., Nam, Y. D. and Kim, H. (2019). Diet Control More Intensively Disturbs Gut Microbiota Than Genetic Background in Wild Type and ob/ob Mice. *Frontiers in microbiology*, 10, 1292. https://doi.org/10.3389/fmicb.2019.01292.

[148] Laursen, M. F., Larsson, M. W., Lind, M. V., Larnkjær, A., Mølgaard, C., Michaelsen, K. F., Bahl, M. I. and Licht, T. R. (2020). Intestinal Enterococcus abundance correlates inversely with excessive weight gain and increased plasma leptin in breastfed infants. *FEMS*

microbiology ecology, 96(5), fiaa066. https://doi.org/10.1093/femsec/fiaa066.

[149] Lemas, D. J., Young, B. E., Baker, P. R., 2nd, Tomczik, A. C., Soderborg, T. K., Hernandez, T. L., de la Houssaye, B. A., Robertson, C. E., Rudolph, M. C., Ir, D., Patinkin, Z. W., Krebs, N. F., Santorico, S. A., Weir, T., Barbour, L. A., Frank, D. N. and Friedman, J. E. (2016). Alterations in human milk leptin and insulin are associated with early changes in the infant intestinal microbiome. *The American journal of clinical nutrition*, 103(5): 1291–1300. https://doi.org/10.3945/ajcn.115.126375.

[150] Spyrou, N., Avgerinos, K. I., Mantzoros, C. S. and Dalamaga, M. (2018). Classic and Novel Adipocytokines at the Intersection of Obesity and Cancer: Diagnostic and Therapeutic Strategies. *Current obesity reports*, 7(4): 260–275. https://doi.org/10.1007/s13679-018-0318-7.

[151] Kazama, K., Usui, T., Okada, M., Hara, Y. and Yamawaki, H. (2012). Omentin plays an anti-inflammatory role through inhibition of TNF-α-induced superoxide production in vascular smooth muscle cells. *European journal of pharmacology*, 686(1-3): 116–123. https://doi.org/10.1016/j.ejphar.2012.04.033.

[152] Sanchis-Chordà, J., Del Pulgar, E., Carrasco-Luna, J., Benítez-Páez, A., Sanz, Y. and Codoñer-Franch, P. (2019). Bifidobacterium pseudocatenulatum CECT 7765 supplementation improves inflammatory status in insulin-resistant obese children. *European journal of nutrition*, 58(7): 2789–2800. https://doi.org/10.1007/s00394-018-1828-5.

[153] Arora, S. and Anubhuti (2006). Role of neuropeptides in appetite regulation and obesity--a review. *Neuropeptides*, 40(6): 375–401. https://doi.org/10.1016/j.npep.2006.07.001.

[154] Wilcox G. (2005). Insulin and insulin resistance. *The Clinical biochemist. Reviews*, 26(2): 19–39.

[155] Simon, M. C., Strassburger, K., Nowotny, B., Kolb, H., Nowotny, P., Burkart, V., Zivehe, F., Hwang, J. H., Stehle, P., Pacini, G., Hartmann, B., Holst, J. J., MacKenzie, C., Bindels, L. B., Martinez,

I., Walter, J., Henrich, B., Schloot, N. C. and Roden, M. (2015). Intake of Lactobacillus reuteri improves incretin and insulin secretion in glucose-tolerant humans: a proof of concept. *Diabetes care*, 38(10): 1827–1834. https://doi.org/10.2337/dc14-2690.

[156] Chang, C. J., Lu, C. C., Lin, C. S., Martel, J., Ko, Y. F., Ojcius, D. M., Wu, T. R., Tsai, Y. H., Yeh, T. S., Lu, J. J., Lai, H. C. and Young, J. D. (2018). Antrodia cinnamomea reduces obesity and modulates the gut microbiota in high-fat diet-fed mice. *International journal of obesity*, 42(2): 231–243. https://doi.org/10.1038/ijo.2017.149.

[157] Weitkunat, K., Stuhlmann, C., Postel, A., Rumberger, S., Fankhänel, M., Woting, A., Petzke, K. J., Gohlke, S., Schulz, T. J., Blaut, M., Klaus, S. and Schumann, S. (2017). Short-chain fatty acids and inulin, but not guar gum, prevent diet-induced obesity and insulin resistance through differential mechanisms in mice. *Scientific reports*, 7(1), 6109. https://doi.org/10.1038/s41598-017-06447-x.

[158] Tuccinardi, D., Farr, O. M., Upadhyay, J., Oussaada, S. M., Klapa, M. I., Candela, M., Rampelli, S., Lehoux, S., Lázaro, I., Sala-Vila, A., Brigidi, P., Cummings, R. D. and Mantzoros, C. S. (2019). Mechanisms underlying the cardiometabolic protective effect of walnut consumption in obese people: A cross-over, randomized, double-blind, controlled inpatient physiology study. *Diabetes, obesity & metabolism*, 21(9): 2086–2095. https://doi.org/10.1111/dom. 13773.

[159] Dao, M. C., Sokolovska, N., Brazeilles, R., Affeldt, S., Pelloux, V., Prifti, E., Chilloux, J., Verger, E. O., Kayser, B. D., Aron-Wisnewsky, J., Ichou, F., Pujos-Guillot, E., Hoyles, L., Juste, C., Doré, J., Dumas, M. E., Rizkalla, S. W., Holmes, B. A., Zucker, J. D., Clément, K. and the MICRO-Obes Consortium (2019). A Data Integration Multi-Omics Approach to Study Calorie Restriction-Induced Changes in Insulin Sensitivity. *Frontiers in physiology*, 9, 1958. https://doi.org/ 10.3389/fphys.2018.01958.

[160] Mukorako, P., Lopez, C., Baraboi, E. D., Roy, M. C., Plamondon, J., Lemoine, N., Biertho, L., Varin, T. V., Marette, A. and Richard, D. (2019). Alterations of Gut Microbiota After Biliopancreatic Diversion

with Duodenal Switch in Wistar Rats. *Obesity surgery*, 29(9): 2831–2842. https://doi.org/10.1007/s11695-019-03911-7.

[161] Shao, Y., Shen, Q., Hua, R., Evers, S. S., He, K. and Yao, Q. (2018). Effects of sleeve gastrectomy on the composition and diurnal oscillation of gut microbiota related to the metabolic improvements. *Surgery for obesity and related diseases: official journal of the American Society for Bariatric Surgery*, 14(6): 731–739. https://doi.org/10.1016/j.soard.2018.02.024.

[162] Cortez, R. V., Petry, T., Caravatto, P., Pessôa, R., Sanabani, S. S., Martinez, M. B., Sarian, T., Salles, J. E., Cohen, R. and Taddei, C. R. (2018). Shifts in intestinal microbiota after duodenal exclusion favor glycemic control and weight loss: a randomized controlled trial. *Surgery for obesity and related diseases: official journal of the American Society for Bariatric Surgery*, 14(11): 1748–1754. https://doi.org/10.1016/j.soard.2018.07.021.

[163] Reijnders, D., Goossens, G. H., Hermes, G., Smidt, H., Zoetendal, E. G. and Blaak, E. E. (2018). Short-Term Microbiota Manipulation and Forearm Substrate Metabolism in Obese Men: A Randomized, Double-Blind, Placebo-Controlled Trial. *Obesity facts*, 11(4): 318–326. https://doi.org/10.1159/000492114.

[164] Olarescu, N. C., Gunawardane, K., Hansen, T. K., Christiansen, J. S. and Jorgensen, J. O. (2000). Normal physiology of growth hormone in adults. In: Feingold, K. R., Anawalt, B., Boyce, A., Chrousos. G., Dungan, K., Grossman, A., Hershman, J. M., Kaltsas, G., Koch, K., Kopp, P., Korbonits, M., McLachlan, R., Morley, J. E., New, M., Perreault, L., Purnell, J., Rebar, R., Singer, F., Trence, D. L., Vinik, A. and Wilson, D. P. (eds.). *Endotext.* South Dartmouth (MA): MD-Text.com. Inc. http://www.ncbi.nlm.nih.gov/books/NBK279056/.

[165] Ohara, E., Tokuyama, H., Kitamoto, T., Kitahara, A., Hayashi, A., Hayashi, H., Takemoto, M. and Yokote, K. (2017). Laparoscopic Sleeve Gastrectomy Resolves Low GHRP-2-Stimulated Growth Hormone Levels in Obese Patients. *Obesity surgery*, 27(8): 2214–2217. https://doi.org/10.1007/s11695-017-2769-4.

[166] Jensen, E. A., Young, J. A., Jackson, Z., Busken, J., List, E. O., Carroll, R. K., Kopchick, J. J., Murphy, E. R. and Berryman, D. E. (2020). Growth Hormone Deficiency and Excess Alter the Gut Microbiome in Adult Male Mice. *Endocrinology*, 161(4), bqaa026. https://doi.org/10.1210/endocr/bqaa026.

[167] Maniam, J. and Morris, M. J. (2010). Voluntary exercise and palatable high-fat diet both improve behavioural profile and stress responses in male rats exposed to early life stress: role of hippocampus. *Psychoneuroendocrinology*, 35(10): 1553–1564. https://doi.org/10. 1016/j.psyneuen.2010.05.012.

[168] Qiu, D., Xia, Z., Deng, J., Jiao, X., Liu, L. and Li, J. (2019). Glucorticoid-induced obesity individuals have distinct signatures of the gut microbiome. *BioFactors (Oxford, England)*, 45(6): 892–901. https://doi.org/10.1002/biof.1565.

[169] Rabiei, S., Hedayati, M., Rashidkhani, B., Saadat, N. and Shakerhossini, R. (2019). The Effects of Synbiotic Supplementation on Body Mass Index, Metabolic and Inflammatory Biomarkers, and Appetite in Patients with Metabolic Syndrome: A Triple-Blind Randomized Controlled Trial. *Journal of dietary supplements*, 16(3): 294–306. https://doi.org/10.1080/19390211.2018.1455788.

[170] Perry, R. J., Peng, L., Barry, N. A., Cline, G. W., Zhang, D., Cardone, R. L., Petersen, K. F., Kibbey, R. G., Goodman, A. L. and Shulman, G. I. (2016). Acetate mediates a microbiome-brain-β-cell axis to promote metabolic syndrome. *Nature*, 534(7606): 213–217. https://doi.org/10.1038/nature18309.

[171] Méndez-Salazar, E. O., Ortiz-López, M. G., Granados-Silvestre, M., Palacios-González, B. and Menjivar, M. (2018). Altered Gut Microbiota and Compositional Changes in *Firmicutes* and *Proteobacteria* in Mexican Undernourished and Obese Children. *Frontiers in microbiology*, 9, 2494. https://doi.org/10.3389/fmicb. 2018.02494.

[172] Acharya, K. D., Gao, X., Bless, E. P., Chen, J. and Tetel, M. J. (2019). Estradiol and high fat diet associate with changes in gut microbiota in

female ob/ob mice. *Scientific reports*, 9(1): 20192. https://doi.org/10.1038/s41598-019-56723-1.

[173] Hassan, A. M., Mancano, G., Kashofer, K., Liebisch, G., Farzi, A., Zenz, G., Claus, S. P. and Holzer, P. (2020). Anhedonia induced by high-fat diet in mice depends on gut microbiota and leptin. *Nutritional neuroscience*, 1–14. Advance online publication. https://doi.org/10.1080/1028415X.2020.1751508.

[174] Nagpal, R., Newman, T. M., Wang, S., Jain, S., Lovato, J. F. and Yadav, H. (2018). Obesity-Linked Gut Microbiome Dysbiosis Associated with Derangements in Gut Permeability and Intestinal Cellular Homeostasis Independent of Diet. *Journal of diabetes research*, 3462092. https://doi.org/10.1155/2018/3462092.

[175] Kayser, B. D., Prifti, E., Lhomme, M., Belda, E., Dao, M. C., Aron-Wisnewsky, J., MICRO-Obes Consortium, Kontush, A., Zucker, J. D., Rizkalla, S. W., Dugail, I. and Clément, K. (2019). Elevated serum ceramides are linked with obesity-associated gut dysbiosis and impaired glucose metabolism. *Metabolomics: Official journal of the Metabolomic Society*, 15(11), 140. https://doi.org/10.1007/s11306-019-1596-0.

[176] Mastrangelo, A., Martos-Moreno, G. Á., García, A., Barrios, V., Rupérez, F. J., Chowen, J. A., Barbas, C. and Argente, J. (2016). Insulin resistance in prepubertal obese children correlates with sex-dependent early onset metabolomic alterations. *International journal of obesity*, 40(10): 1494–1502. https://doi.org/10.1038/ijo.2016.92.

[177] Sheikh, A. B., Nasrullah, A., Haq, S., Akhtar, A., Ghazanfar, H., Nasir, A., Afzal, R. M., Bukhari, M. M., Chaudhary, A. Y. and Naqvi, S. W. (2017). The Interplay of Genetics and Environmental Factors in the Development of Obesity. *Cureus*, 9(7), e1435. https://doi.org/10.7759/cureus.1435.

[178] Conlon, M. A. and Bird, A. R. (2014). The impact of diet and lifestyle on gut microbiota and human health. *Nutrients*, 7(1): 17–44. https://doi.org/10.3390/nu7010017.

BIOGRAPHICAL SKETCHES

Paula Robles Bolívar

Affiliation: Biochemistry and Molecular Biology I, School of Sciences, University of Granada, Granada, Spain

Education: Bachelor's degree in biochemistry, Master´s programme in Genetics and Evolution

Business Address: Institute of Nutrition and Food Technology "José Mataix", Center of Biomedical Research, University of Granada, Avda. del Conocimiento s/n., 18016 Armilla, Granada, Spain.

Research and Professional Experience:

- 2014-2016: Postgraduate researcher in Department of Biochemistry, Cellular and Molecular Biology of Plants in Zaidin Experimental Station (EEZ), Centre of the Spanish Council for Scientific Research (CSIC). Granada, Spain.
- 2016-2017: Pre-doctoral researcher in Department of Biochemistry in Rovira and Virgili University, Tarragona, Spain. Evaluating the enteroendocrine and immunoprotective effects of flavanols on the gastrointestinal wall.
- 2017-2019: Superior laboratory technician in Era7Bioinformatics company. NGS libraries preparation and sequencing on Illumina platforms.
- 2020: Superior laboratory technician in Department of Biochemistry and Molecular Biology I, School of Sciences, University of Granada, Granada, Spain

Publications from the Last 3 Years:

1) Jimenez-Lopez JC, Robles-Bolivar P, Lopez-Valverde FJ, Lima-Cabello E, Kotchoni SO, Alché JD "Ole e 13 is the unique food allergen

in olive: structure-functional, substrates docking, and molecular allergenicity comparative analysis" *Journal of Molecular Graphics and Modelling.* 66, 26- 40 (2016).

2) Jimenez-Lopez JC, Lopez-Valverde FJ, Robles-Bolivar P, Lima-Cabello E, Gachomo EW, Kotchoni SO "Genome-Wide Identification and Functional Classification of Tomato (Solanum lycopersicum) Aldehyde Dehydrogenase (ALDH) Gene Superfamily" *PLOS ONE* (2016) doi: 10.1371/journal.pone.0164798.

3) Lima-Cabello E, Robles-Bolivar P, Alché J.D, Jimenez-Lopez JC "Narrow Leafed Lupin Beta- Conglutin Proteins Epitopes Identification and Molecular Features Analysis Involved in Cross- Allergenicity to Peanut and Other Legumes" *Genomics and Computational Biology.* 2(1), e29 (2016).

4) Ginés I, Gil-Cardoso K, Robles P, Arola L, Terra X, Blay M, Ardévol A, Pinent M. "A novel ex vivo experimen-tal setup to investigate how food components stimulate the enteroendocrine secretions of different intestinal segments." *Journal of Agricultural and Food Chemistry.* doi: 10.1021/acs.jafc.8b03046 (2018).

Ana Isabel Álvarez- Mercado

Affiliation:

- Department of Biochemistry and Molecular Biology II, School of Pharmacy, University of Granada, 18071 Granada, Spain.
- Institute of Nutrition and Food Technology "José Mataix," Center of Biomedical Research, University of Granada, Avda. del Conocimiento s/n., 18016 Armilla, Granada, Spain.
- Instituto de Investigación Biosanitaria IBS.GRANADA, Complejo Hospitalario Universitario de Granada, 18014 Granada, Spain.

Education:

- Name of qualification: PhD Chemistry. Year: 2009. Degree awarding entity: University of Granada

- Name of qualification: Molecular and Cellular Immunology. Degree awarding entity: University of Granada. Year: 2007
- Name of qualification: Degree in Chemistry. Degree awarding entity: University of Jaen. Year: 2002

Business Address: Institute of Nutrition and Food Technology "José Mataix," Center of Biomedical Research, University of Granada, Avda. del Conocimiento s/n., 18016 Armilla, Granada, Spain.

Research and Professional Experience:
- 2018-present; Postdoctoral Researcher, Department of Biochemistry and Molecular Biology. University of Granada. Spain.
- 2016-2018; Postdoctoral Researcher, Institut d'investigacions Biomèdiques August Pi i Sunyer, Barcelona. Spain.
- 2015; Researcher, Valle Hebrón Hospital. Barcelona. Spain.
- 2014; Associate Researcher, King's College London. London, United Kingdom
- 2013; Postdoctoral Fellow, Northwestern University. Chicago U.S.
- 2011-2012; Postdoctoral Fellow, Andalusian Center for Molecular Biology and Regenerative Medicine. Seville. Spain.
- 2007-2011; Researcher, Department of Biochemistry and Molecular Biology. University of Granada. Spain.

Publications from the Last 3 Years:
- Solis-Urra P, Plaza-Diaz J, Álvarez-Mercado AI, Rodríguez-Rodríguez F, Cristi-Montero C, Zavala-Crichton JP, Olivares-Arancibia J, Sanchez-Martinez J, Abadía-Molina F. The Mediation Effect of Self-Report Physical Activity Patterns in the Relationship between Educational Level and Cognitive Impairment in Elderly: A Cross-Sectional Analysis of Chilean Health National Survey 2016-2017. *Int J Environ Res Public Health.* 2020; 17(8). pii: E2619. doi: 10.3390/ijerph17082619.

- Cornide-Petronio ME, Álvarez-Mercado AI, Jiménez-Castro MB, Peralta C. Current Knowledge about the Effect of Nutritional Status, Supplemented Nutrition Diet, and Gut Microbiota on Hepatic Ischemia-Reperfusion and Regeneration in Liver Surgery. *Nutrients.* 2020; 12(2). pii: E284. doi: 10.3390/nu12020284.

- Álvarez-Mercado AI, Negrete-Sánchez E, Gulfo J, Ávalos de León CG, Casillas-Ramírez A, Cornide-Petronio ME, Bujaldon E, Rotondo F, Gracia-Sancho J, Jiménez-Castro MB, Peralta C. EGF-GH Axis in Rat Steatotic and Non-steatotic Liver Transplantation From Brain-dead Donors. *Transplantation.* 2019; 103(7):1349-1359. doi: 10.1097/TP.0000000000002636.

- Tobar HE, Cataldo LR, González T, Rodríguez R, Serrano V, Arteaga A, Álvarez-Mercado A, Lagos CF, Vicuña L, Miranda JP, Pereira A, Bravo C, Aguilera CM, Eyheramendy S, Uauy R, Martínez Á, Gil Á, Francone O, Rigotti A, Santos JL. Identification and functional analysis of missense mutations in the lecithin cholesterol acyltransferase gene in a Chilean patient with hypoalphalipoproteinemia. *Lipids Health Dis.* 2019; 18(1):132. doi: 10.1186/s12944-019-1045-0.

- Plaza-Díaz J, Álvarez-Mercado AI, Ruiz-Marín CM, Reina-Pérez I, Pérez-Alonso AJ, Sánchez-Andujar MB, Torné P, Gallart-Aragón T, Sánchez-Barrón MT, Reyes Lartategui S, García F, Chueca N, Moreno-Delgado A, Torres-Martínez K, Sáez-Lara MJ, Robles-Sánchez C, Fernández MF, Fontana L. Association of breast and gut microbiota dysbiosis and the risk of breast cancer: a case-control clinical study. *BMC Cancer.* 2019; 19(1):495. doi: 10.1186/s12885-019-5660-y.

- Álvarez-Mercado AI, Navarro-Oliveros M, Robles-Sánchez C, Plaza-Díaz J, Sáez-Lara MJ, Muñoz-Quezada S, Fontana L, Abadía-Molina F. Microbial Population Changes and Their Relationship with Human Health and Disease. *Microorganisms.* 2019; 7(3). pii: E68. doi: 10.3390/microorganisms7030068.

- Álvarez-Mercado AI, Gulfo J, Romero Gómez M, Jiménez-Castro MB, Gracia-Sancho J, Peralta C. Use of Steatotic Grafts in Liver

Transplantation: Current Status. *Liver Transpl.* 2019; 25(5):771-786. doi: 10.1002/lt.25430.

- Cornide-Petronio ME, Bujaldon E, Mendes-Braz M, Avalos de León CG, Jiménez-Castro MB, Álvarez-Mercado AI, Gracia-Sancho J, Rodés J, Peralta C. J The impact of cortisol in steatotic and non-steatotic liver surgery. *Cell Mol Med.* 2017; 21(10):2344-2358. doi: 10.1111/jcmm.13156.

- Jiménez-Castro MB, Negrete-Sánchez E, Casillas-Ramírez A, Gulfo J, Álvarez-Mercado AI, Cornide-Petronio ME, Gracia-Sancho J, Rodés J, Peralta C.The effect of cortisol in rat steatotic and non-steatotic liver transplantation from brain-dead donors. *Clin Sci* (Lond). 2017 25; 131(8):733-746. doi: 10.1042/CS20160676.

In: Dysbiosis: A Study of Underlying Causes   ISBN: 978-1-53618-332-0
Editor: Richard I. Cowell   © 2020 Nova Science Publishers, Inc.

Chapter 2

INSIGHTS INTO THE ROLE OF DYSBIOSIS IN THE PROGRESSION OF LIVER DISEASE: AN UPDATE

Teresa Rubio-Tomás[1,2], Ascensión Rueda-Robles[3,4]
and Ana I. Álvarez-Mercado[3,4,5,*]

[1]Institut d'Investigacions Biomèdiques August Pi i Sunyer (IDIBAPS),
Barcelona, Spain
[2]School of Medicine, University of Crete, Herakleion, Crete, Greece
[3]Department of Biochemistry and Molecular Biology II,
School of Pharmacy, University of Granada, Granada, Spain
[4]Institute of Nutrition and Food Technology "José Mataix,"
Biomedical Research Center, Granada, Spain
[5]Instituto de Investigación Biosanitaria ibs. GRANADA,
Granada, Spain

---

[*] Corresponding Author's Email: alvarezmercado@ugr.es.

ABSTRACT

The liver is the central organ for metabolic processes including xenobiotic metabolism. An unhealthy status usually presents an altered gut microbiota profile (known as dysbiosis) associated with the loss of its functions.

The communication (gut-liver axis) between the liver and the gut is bidirectional and is mediated by the biliary tract, the portal vein, and the systemic circulation, with bile acids being key mediators in this communication. The influence of dysbiosis on the prevalence and pathogenesis of liver diseases is critical, ranging from simple steatosis (defined as excessive accumulation of triglycerides in hepatocytes) and steatohepatitis to more severe complications such as fibrosis, advanced cirrhosis and hepatocellular carcinoma. Additionally, taking into account the complex relationship among the gut microbiome, immune cells, and tumor cells in the liver, the characterization of alterations in the gut microbiota and their underlying causes related to the aforementioned diseases could have a great impact on the diagnosis, prevention, and design of potential therapeutic strategies. Knowledge in this regard could contribute to detaining the progression of disease from simple steatosis to hepatic cancer.

This review summarizes and critically discusses recent literature on the evaluation of changes in the microbiota associated with liver diseases. We aim to evaluate which alterations could be involved in the initiation and progression of liver disease.

Keywords: gut-liver axis, gut microbiota, dysbiosis, alcoholic liver disease, non-alcoholic liver disease, non-alcoholic steatohepatitis, hepatocellular carcinoma

INTRODUCTION

The liver is the main organ involved in metabolic processes including xenobiotic metabolism, and it has a bidirectional relationship with the gut and the microbiota, which is known as the gut-liver axis. This relationship includes reciprocal cellular and molecular interactions produced and affected by dietary, genetic and environmental factors [1]. The portal vein allows the transport of nutrients and bacterial compounds and their metabolites from the intestinal lumen through the gut barrier to the liver, contributing to homeostasis under healthy physiological conditions [2]. Failures and/or alterations of this beneficial interaction result in dysbiosis (loss of balance in microbiota composition and function) [3], inducing disturbances in the intestinal barrier. As a result, an increment in the portal influx of bacteria or their products to the liver is produced [4].

Overall, the bacterial microbiota is mostly made up of the phyla *Firmicutes* (64%), with more than 250 genera, and *Bacteroidetes* (23%), including genera such as *Bacteroides, Flavobacteria, Sphingobacteriales,* as well as *Bacteroides thetaiotaomicron* (species), *Proteobacteria* (8%) and *Actinobacteria* (3%) (phyla) [5]. In a healthy human gut, the most characteristic bacterial phyla are *Firmicutes, Bacteroides, Proteobacteria, Actinobacteria, Fusobacteria* and *Verrucomicrobia* [6, 7]. Alterations in the gut microbiota are important factors in the occurrence and progression of several liver diseases such as alcoholic liver disease (ALD) and non-alcoholic liver disease (NAFLD) [8], both considered a major health burden in industrialized countries [9]. Indeed, the progressive adoption of the Western lifestyle, including changes in nutritional habits and alcohol abuse, has led to an increment in the incidence and prevalence of NAFLD and related metabolic disorders as well as ALD, respectively [8, 10]. Furthermore, ALD and NAFLD encompass a broad spectrum of hepatic lesions from asymptomatic steatosis to more severe complications such as steatohepatitis, fibrosis, cirrhosis, and hepatocellular carcinoma (HCC) [11], being end-stage liver disease one of the most common causes of morbidity and mortality worldwide [12].

The susceptibility of alcoholic patients to the development of ALD can very widely among individuals, and its progression to more advanced stages is highly influenced by several factors such as the amount and duration of alcohol abuse, and also the gut microbiota and its metabolites [11]. A large percentage of patients with a history of alcohol abuse present changes in the microbiota, similar to what has been observed in studies performed in animals [13-22]. Moreover, alcohol-induced changes in endogenous and gut microbiota-derived metabolites are associated with a specific metabolomic profile [23-25].

The pathophysiology of NAFLD has been linked to lower microbial diversity and a weakened intestinal barrier, exposing the host to bacterial components and stimulating pathways of immune defense and inflammation [3]. In addition, disrupted host-microbial metabolic interaction alters bile acid signaling. In the last few years, it has been demonstrated that the gut microbial composition of patients with NAFLD or non-alcoholic steatohepatitis (NASH) differs from that of healthy individuals with variable degrees of alterations [2]. These alterations are more important in cirrhosis, increasing the risk of clinically significant portal hypertension and leading to bacterial translocation, sepsis and acute-on-chronic liver failure [3]. Consistent microbiome signatures discriminating healthy individuals from those with NAFLD, NASH or cirrhosis have been reported [26].

Taking into account all of the above, recent evidence have established a key role of the gut microbiota in the pathogenesis of the broad spectrum of liver diseases [8, 11] although the underlying mechanisms are not yet understood in depth. Therefore, the gut microbiota and its bacterial products may contribute to the development of liver diseases and their progression through mechanisms such as increased intestinal permeability, chronic systemic inflammation, or changes in the metabolism, among others. This review aims to highlight and discuss the literature published during the last 5 years on the identification of changes in the gut microbiota related to liver diseases.

ALCOHOLIC LIVER DISEASE

Pathophysiology

Alcohol metabolism involves mainly the liver but also, to a lesser extent, the gastrointestinal tract, which explains the great degree of tissue injury induced by heavy drinking [11]. Up to 80-90% of patients with chronic heavy alcohol intake develop steatosis, and 10–20% finally progress to cirrhosis [27]. Progression of ALD is affected by the amount and duration of alcohol abuse, as well as genetic and epigenetic factors. The microbiota and its metabolites have been shown to play a role in the physiopathology of ALD [11]. Alcohol overconsumption leads to gut dysbiosis by an increase in potentially pathological bacteria and a decrease in beneficial bacterial species as well as their microbial products. Additionally, alcohol overuse causes disruption of the intestinal barrier, provoking the translocation of bacterial components, viable bacteria and their metabolites (acetaldehyde being the most important) to the circulatory system [13]. In long-term alcohol consumers, gut-derived endotoxins cause the translocation of specific microbe-associated molecular patterns such as lipopolysaccharide (LPS) into the systemic circulation [2] and promote progression from simple steatosis to inflammation and fibrosis [10].

"Healthy" microbiota is thought to protect against alcohol-induced liver injury [28] and, even in the absence of liver disease, alcohol abuse and acute binge drinking can produce dysbiosis [29]. In addition, not only gut bacteria but also fungi seem to play a role in ALD progression [30]. Systemic immune response to fungi or fungal products is associated with increased mortality in alcoholic hepatitis [31]. In this regard, *Candida* is the most abundant commensal gut fungus genus in alcohol disorders being *Candida albicans* exotoxin (*candidalysin*) associated with the severity of and mortality by alcoholic hepatitis. *Candidalysin* probably promotes ALD, inducing hepatocyte damage [32].

Remarkably, data about the molecular pathways regulating the effects of ethanol are scarce and, unfortunately, the existing animal models do not completely replicate the pathogenesis, and clinical and histological features observed in ALD patients, and often do not exactly match a specific stage of ALD. The composition of the microbiota in patients with alcoholic hepatitis with underlying cirrhosis is different from that of cirrhotic patients, and presents specific changes in bile acid homeostasis. On one hand, changes in the bile acid pool promote an alteration of the gut microbiota composition and, on the other hand, the microbiota participates in bile acid metabolism. Thus, although more studies are needed to define the contribution of the gut microbiota to the pathogenesis of ALD and alcoholic hepatitis, it seems to be a complicated vicious cycle with many key players that worsen the pathological situation [33].

Studies on ALD

The Lieber-DeCarli model of chronic *ad libitum* ethanol administration [34], the chronic and binge ethanol feeding model (Bin Gao-National Institute on Alcohol Abuse and Alcoholism (NIAAA) model) [35] and the Tsukamoto–French model of chronic intragastric ethanol administration [36] are widely accepted, in spite of their limitations (i.e., different degrees of liver injury, steatosis and inflammation) [37-39]. Specific experimental models show specific effects of alcohol consumption. For example, alcohol-preferring rats displayed hepatic steatosis and increased serum transaminases and LPS concentrations, as well as changes in gut microbiota composition after 3-12 months of voluntarily consuming alcohol, but did not show liver necrosis or inflammation, systemic inflammation or alteration of the intestinal colonic mucosa, meaning that their gut barrier was intact [40]. In mice, specific drinking patterns, such as the chronic intermittent consumption paradigm, but not the "drinking in the dark" binge model [41], promoted bacterial translocation and increased LPS plasma levels, as well as decreased expression of zonula occludens 1 (ZO-1) and occludin, and enhanced matrix metalloproteinase 9 activity in the colon. Furthermore, a chronic intermittent consumption paradigm might affect brain tryptophan metabolism by an unknown mechanism involving the gut microbiota [38].

It has been reported that alcohol and circadian rhythms are mutually regulated by each other. On one hand, genetic (by mutation on Clock) and environmental disruption (by shifts in the light/darkness cycle) of circadian rhythms in mice resulted in gut leakiness, and when combined with alcohol consumption, promoted alcohol-induced gut leakiness, endotoxemia (increased serum LPS levels) and steatosis and inflammation in the liver (steatohepatitis) [42]. On the other hand, alcohol-consuming rats presented gut leakiness and increased PER2 circadian clock protein expression in the duodenum and colon [43].

Interestingly, most of the effects of acute alcohol intake (i.e., binge drinking) seemed to be reversible and did not necessarily lead to ALD. Indeed, a single alcohol binge in healthy young volunteers transiently affected neutrophil functionality. Neutrophils isolated from the blood of healthy volunteers up to 4 hours or the day after a single alcohol binge showed an enhanced oxidative burst function, accompanied by an elevated percentage of neutrophils that presented impaired phagocytic capacity. In contrast, microbiota composition, gut permeability, bacterial translocation and inflammation were not affected in these conditions [44]. To the contrary, even low-dose short-term (1 week) consumption of alcohol altered gut microbiota composition in mice. These changes were apparent immediately (after 1 and 7 days of alcohol drinking) and were reversed by supplementation with microbes contained in fermented rice liquors, which increased the microbial production of short-chain fatty acids (SCFAs) such as butyric acid and propionic acid, and suppressed intestinal inflammation [45]. Additionally, *in vitro* stimulation of enterocytes with microbial products from chronic ethanol-fed mice led to increased intestinal permeability, while stimulation of peripheral blood mononuclear cells with microbial products resulted in their activation [46].

Molecular Pathways Involved in ALD Pathogenesis through the Gut Microbiota

Hypoxia-inducible factor 1α (HIF-1α) may also mediate ALD by modulating the gut microbiota. Mice devoid of HIF-1α in the intestinal epithelium displayed worsening of the effects of chronic alcohol consumption. These effects included liver steatosis and injury, elevated serum LPS levels, decreases in the expression of intestinal epithelial tight junction proteins claudin-1 and occludin (that are related to increased intestinal permeability), intestinal inflammation (mediated by macrophages and pro-inflammatory cytokines), and dysbiosis. Besides HIF-1α, another interesting molecule is immunoglobulin A, which binds bacteria in the intestine and hampers bacterial translocation. Unexpectedly, the development of ALD was not affected in *Iga (-/-)* mice subjected to chronic ethanol feeding, probably due to increased compensatory secretion of immunoglobulin M by the intestine [47]. Moreover, a role for purinergic P2X7 receptor in alcohol-induced steatohepatitis and intestinal injury has been proposed. P2X7 receptor antagonists decreased alcohol-induced liver and intestinal injury and inflammation by inhibiting the mitogen-activated protein kinase extracellular signal in a murine model of chronic plus binge alcohol feeding [48]. The physiopathological relevance of the LPS-Toll-like receptor-4 (TLR4) axis in macrophages in ALD was explored in a study feeding mice with a chronic ethanol diet (6 weeks) supplemented with the dietary non-absorbable fiber inulin. Inulin modulated gut microbiota composition and reversed ethanol-induced increases in serum transaminases and inflammatory factors, thereby ameliorating ALD. These effects were concomitant to a reduction in TLR4-expressing macrophages, suggesting a molecular mechanism involving the LPS-TLR4 pathway [49]. Antimicrobial-regenerating islet-derived (REG)-3 lectins have also been studied in the context of ALD and dysbiosis. Intestine-specific *Reg3b* or *Reg3g* knockout mice presented exacerbated alcohol-induced liver disease (steatohepatitis), as well as higher mucosa-associated bacteria load and bacterial translocation to the liver and mesenteric lymph nodes, suggesting that dysbiosis might be the mechanism that mediates the progression of alcoholic fatty liver disease to steatohepatitis. On the contrary, overexpression of *Reg3g* in intestinal epithelial cells reversed these effects. This study points to the role of alcohol in the dysregulation of mucosa-associated microbiota, which led to mucosal barrier dysfunction and facilitated ALD progression [50]. A-defensin, the main antimicrobial peptide synthesized and secreted by Paneth cells in the intestine, has also been proposed as a link between gut dysbiosis and ALD. Genetic α-defensin dysfunction exacerbated chronic ethanol consumption (8 weeks)-induced pathogen-associated molecular patterns (PAMPs) translocation and liver injury. Oral consumption of synthetic human α-defensin 5 reversed these effects. Importantly, dietary zinc deficiency may be partially responsible for Paneth cell α-defensin dysfunction. This mechanism was postulated to play a role in ALD development and alcoholic hepatitis, in which PAMP translocation has been observed [51]. *Enterococcus faecalis* exotoxin cytolysin may also be responsible for hepatocyte death and liver injury in alcoholic hepatitis [52].

Despite these advances, molecular mechanisms mediating the effects of alcohol on the microbiota are still unclear.

Supplemental Approaches to Ameliorate the Effects of Alcohol Consumption by Modulation of Gut Microbiota

A wide range of substances has been reported to mitigate the effects of alcohol consumption. For example, supplementation with the antioxidant butyrate in the form of triglyceride tributyrin partially reversed ethanol-induced oxidative stress and innate immune responses in the colon of mice exposed to a chronic-binge ethanol diet [53]. Other interventions in this regard are described below.

Plant-Derived Supplements

Supplementation with rhubarb extract promoted enrichment of *Akkermansia muciniphila* and *Parabacteroides goldsteinii* in the intestine and ameliorated alcohol-induced hepatic injury by attenuating oxidative stress and inflammation in mice subjected to a single alcohol binge. However, this association does not necessarily imply that these bacterial genera were involved in the improvement of the hepatic injury phenotype [54].

Kaempferol (a plant-derived flavonoid) supplementation previous to a single ethanol binge protected against ethanol-induced increases in serum transaminases and restored the expression of tight junction proteins ZO-1 and occludin, as well as butyrate receptor GPR109A and butyrate transporter SLC58A in the ileum and colon [55]. Other substances that have recently been proposed to present therapeutic properties by reversing ethanol-induced pathological effects through modulation of the microbiota in murine models include: water extracts from mulberry and white flower dandelion for liver steatosis [56], *Decaisnea insignis* seed oil for liver injury and metabolic alterations [57], garlic polysaccharide for liver fibrosis [58] and the carotenoid astaxanthin for liver injury and steatosis in mice [59].

Probiotics and Antibiotics

Synbiotic supplementation (i.e., combining probiotics and prebiotics) protected against the ethanol-induced reduction of gut microbiota abundance and diversity and against the disruption of hepatocyte and liver sinusoidal endothelium barrier integrity in a murine model of short chronic (10 days) plus binge ethanol exposure, suggesting that modulation of gut microbiota composition improves liver barrier integrity, although the mechanism is not clear [60]. Additionally, treatment with antibiotics reduced alcohol-induced pro-inflammatory cytokine expression in the small intestine and the brain and also reversed the alcohol-induced increase in neutrophil infiltration, inflammatory factors and hepatic steatosis. Nevertheless, it did not suppress the alcohol-induced increment in serum transaminases in short chronic (10 days) plus binge ethanol-exposed mice [61, 62], further confirming the well-known role of the gut microbiota in inflammation [38].

Despite the aforementioned benefits, another study found that germ-free mice are more susceptible to single alcohol binge-induced liver injury, thereby highlighting the importance of a healthy microbiota in providing protection against liver insults [28]. However, controversial results indicate that germ-free status can be associated either with an exaggerated or a protective phenotype depending on the liver disease [14].

NON-ALCOHOLIC FATTY LIVER DISEASE

Pathophysiology

NAFLD affects 25% of the global adult population and is considered one of the most prevalent chronic liver diseases worldwide [63]. However, not only adults but also obese children and adolescents suffer from this disorder [64]. The main feature of NAFLD is lipid accumulation within hepatocytes >5% of the total liver weight (especially in the perivenular region and periportal areas) [65]. Male sex, age, metabolic syndrome, insulin resistance and alanine aminotransferases are associated with NAFLD, the major risk factors being the presence of obesity and type 2 diabetes [66, 67].

Additionally, other factors, such as genetic predisposition, sedentary lifestyle and hypercaloric diet, are associated with the development of NAFLD [5]. In this regard, the gut microbiota is also considered critical for the development of NAFLD [5, 68, 69]. Indeed, it has been reported that NAFLD patients show fewer numbers of *Bacteroidetes* and higher numbers of *Prevotella* and *Porphyromas* than healthy subjects as well as elevated concentrations of *Lactobacillus* (*L.*), *Escherichia* (*E.*) and *Streptococcus* and diminished levels of *Ruminococcaceae* and *Faecalibacterium prausnitzii* [70]. Other authors have also described an increase in the *Firmicutes/Bacteroidetes* (F/B) ratio in NAFLD patients [71, 72]. This increment may lead to elevated activity of Toll-like receptors (TLRs) and nucleotide oligomerization domain pathways, which could result in an alteration of the tight junction protein leading to increased gut permeability. The latter leads to an augmentation of hepatic expression of pro-inflammatory cytokines due to accumulation of toxic bacterial products such as LPS [73].

The Interplay between Diet and the Gut Microbiota in NAFLD

It is known that diet is an important modulator of the gut microbiota. For instance, an imbalance in the omega-3 and omega-6 polyunsaturated essential fatty acids ratio (n-3/n-6 PUFA) can lead to inflammation, overweight or insulin resistance, which are factors intimately associated with NAFLD [73]. In addition, *de novo* hepatic lipogenesis is stimulated by high levels of SCFAs and low concentrations of adiponectin in visceral adipose tissue, which leads to the production of intrahepatic molecules such as diacylglycerols and ceramides, and promotes the emergence of insulin resistance and even the activation of Kupffer and hepatic stellate cells (HSCs) resulting in necro-inflammation [74]. A high-fat diet (HFD) can also modulate the composition of the gut microbiota by decreasing protective intestinal bacteria and favoring the prevalence of opportunistic pathogenic products of Gram-negative bacteria, such as LPS [75]. HDF favors the growth of *Bilophila wadsworthia*, which produces hydrogen sulfide, promotes inflammation and affects the intestinal barrier. This contributes to an increase in the release of LPS and decreases the production of butyrate [76]. The presence of LPS in portal circulation enables the binding to TLR4 and other co-receptors in the liver associated with inflammation [68]. *Helicobacter pylori* and *Klebsiella pneumoniae* also contribute to the development of NAFLD. The first promotes the passage of bacterial endotoxins and participates in the release of pro-inflammatory cytokines such as tumor necrosis factor α (TNF-α) and interleukins (IL) such as IL-1β, IL-6 and IL-8 (which also contribute to hepatocellular injury) [76], while *Klebsiella pneumoniae* produces huge levels of endogenous alcohol [77], and an increase in the production of liver reactive oxygen species (ROS) [78]. Conversely, the presence of *Akkermansia muciniphila* reinforces the intestinal barrier, resulting in an improvement in metabolic disorders, lowering cholesterol levels and attenuating liver steatosis [76]. *Allobaculum* has been associated with an improved intestinal barrier, reduced hepatic steatosis/inflammation and better glucose tolerance and metabolism. Accordingly, a diet rich in inulin (8%) was able to restore the effects of the environmental pollutant PCB 126 on the murine intestinal microbiota by an increment of *Allobaculum* and a decrease of *Coprococcus*. Inulin also produced a significant increase of *Bifidobacterium (B.)* and *Lactobacillus (L.)* and reduced the abundance of *Ruminococcus*. These effects were associated with the recovery of homeostasis in the gut microbiota [79].

Notably, there is wide variability in the data extracted from diet-induced obesity NAFLD studies. For instance, a study reported by Porras and collaborators, showed a significant increase in plasma triglyceride and alanine aminotransferase in mice recieving a HFD [80], whereas no differences were found in the study carried out by Liu and collaborators [81]. Chen and collaborators assessed the effects of different housing conditions in germ-free C57BL/6JNarl mice subjected to HFD and gentamicin treatment. The analysis of gut microbiota composition using the 16S rRNA genes indicated no differences between groups fed with HFD (with or without gentamicin). Concerning phyla, *Firmicutes* was a characteristic in specific pathogen-free microbiota, especially when animals were treated with gentamicin, whereas *Bacteroidetes* dominated animals kept in conventional housing. The authors concluded that the gut microbiota varies with housing conditions, suggesting that this could be a more important factor than diet in shaping gut microbiota composition and could explain the differences found in other studies [82].

Nutritional Supplementation and Functional Foods as Drivers to Modulate the Microbiota in NAFLD

Antioxidants

Flavonoids have beneficial effects on lipid metabolism, insulin resistance, inflammation and oxidative stress. Quercetin supplementation decreased insulin resistance in a mouse model of obesity, metabolic syndrome and hepatic steatosis induced by HFD. Diet produced an increase in the F/B ratio and Gram-negative bacteria, as well as an increase in the presence of *Helicobacter* genus. As a result, quercetin reverted the gut microbiota imbalance and related endotoxemia-mediated TLR4 pathway induction [80].

Some authors have suggested the involvement of flavonoids in peroxisome proliferator-activated receptor (PPAR) activity. PPARs are important in the treatment of NAFLD since they decrease steatosis by stimulation of β-oxidation and inhibition of nuclear factor-κB (NF-κB) [83, 84].

Resveratrol can ameliorate HFD-induced hepatic steatosis by the modulation of the gut microbiota. Supplementation with resveratrol plus fecal microbiota transplant (FMT) in mice produced an increment of *Bacteroidetes* as well as a reduction of *Firmicutes* and *Proteobacteria* associated with the inhibition of lipogenesis [85].

Vitamin E has also been proposed as an alternative for the treatment of NAFLD but the results to date are still unclear. Overall, the authors found significant biochemical and histological improvements in adults, whereas only one study carried out in children showed significant improvements in liver function [86].

Sugar

Results from animals and humans revealed that diets high in sugar increase the risk for NAFLD, which is linked to the metabolism of fructose [87]. In addition, an increase of intestinal permeability contributed to stimulating the absorption of monosaccharides leading to bacterial translocation, the production of endotoxins and an increment of lipoprotein lipase activity. Thereafter, a decrease of fasting-induced adipose factor promoted *de novo* synthesis of fatty acids, triglyceride production, and activated inflammatory TLRs in hepatocytes [5]. Moreover, deficiencies of carbohydrate-responsive-element-binding protein reduced hepatic fat levels in a mouse model, which could be an interesting mechanism to be considered [88].

Choline

Choline levels also appear to affect the composition of the intestinal microbiota. Choline deficiency increased Gram-negative bacteria and *Erysipelotrichi* levels, which have been associated with an increment in liver fat content [89]. In a study carried out under anaerobic conditions, *Proteobacteria, Firmicutes*, and *Actinobacteria* catalyzed the conversion of choline to trimethylamine, and its conversion in the liver to trimethylamine-N-oxide, which is high in NAFLD patients [90], worsening or leading to progression in liver steatosis by the inhibition of Farnesoid X receptor (FXR) [91].

Bacterial Metabolites Derived from Aromatic Amino Acids and Branched-Chain Amino Acids

Bacterial metabolites derived from aromatic amino acids have also been related to NAFLD development. The compounds indole, indole-3-propionic acid, indole-3-acetic acid, indole-3-aldehyde, tryptamine, and 3-methylindole maintain intestinal integrity, reduce bacterial translocation, prevent the release of microbiota-derived components and limit inflammatory cascades [92]. Liu and collaborators indicated that patients with NAFLD have an increased abundance of Lachnospiraceae/

Veillonellaceae (phylum *Firmicutes*) and *Kiloniellaceae/Pasteurellaceae* (phylum *Proteobacteria*) [21].

Branched-chain amino acids could also be used as an alternative to prevent the progression of NAFLD since these compounds mitigate hepatic steatosis and liver injury by suppressing fatty acid synthase gene expression and protein levels [93].

Probiotics

Probiotics have shown a positive effect on the production of pro- and anti-inflammatory cytokines. For instance, an increase in insulin sensitivity and a reduction of fat liver accumulation was observed after intervention with *Bifidobacteria* [78]. A double-blinded, placebo-controlled trial carried out in obese NAFLD patients using a probiotic mixture of *L. acidophilus, L. rhamnosus, L. paracasei, Pediococcus pentosaceus, B. lactis,* and *B. breve* for 12 weeks revealed a reduction of intrahepatic fat associated with body weight loss in NAFLD patients [94]. Duseja and collaborators also analyzed the effect of multi-strain probiotics in patients with NAFLD plus lifestyle modification for 1 year showing beneficial effects in liver histology, alanine aminotransferases and cytokines [95].

NON-ALCOHOLIC STEATOHEPATITIS

Pathophysiology

The global prevalence of NASH is approximately 5-7% [96]. Although the mechanism whereby certain patients develop NASH is unclear, some theories have been proposed. The first is the "Two-hit" theory in which the first hit would be the accumulation of fat in the liver (steatosis), whereas the passage from steatosis to steatohepatitis (second hit) occurs by lipid peroxidation and ROS production [77]. The second and most recent theory is the "Multiple-hit" theory, which suggests that inflammation precedes steatosis in environmentally and genetically predisposed subjects [78].

Among others, an increase of intestinal permeability may be related to inflammatory changes in NASH, since there is a strong association between these two processes [97]. In the presence of chronic NAFLD, lobular inflammation and signs of hepatocellular damage can occur, resulting in NASH. In fact, this disease can be described as a necro-inflammatory complication of persistent hepatic steatosis [98]. Additionally, in patients with NASH, a higher prevalence of small intestinal bacterial overgrowth [99] has been related to a higher TNF-α level [100]. This has been linked with increased expression of TLR4 on CD14-positive monocytes and higher plasma IL-8 levels [101]. However, contradictory results have been reported in this sense; i.e., a cross-sectional study performed by Fitriakusumah and collaborators did not show an association among small intestinal bacterial overgrowth, hepatic steatosis and fibrosis in NAFLD patients. [102].

Additionally, gut microbiota profiles can provide information about the degree of severity of NAFLD. Higher concentrations of *Bacteroidetes* are independently and positively associated with NASH while *Ruminococcus* concentration is an independent predictor of fibrosis stage. In addition, *Prevotella* levels are significantly reduced in NASH patients with fibrosis [103]. An increase of *Ruminococcu* s and *Dorea* has been associated with the progression of NASH in pediatric patients [104]. In addition, Duarte and collaborators found decreased concentrations of *Faecalibacterium* in lean NASH patients and concluded that a reduction in the production of SCFAs may induce a pro-inflammatory environment promoting the progression of NASH [70]. Accordingly, *Faecalibacterium* produces molecules blocking NF-kB activation and IL-8 secretion, which exert anti-inflammatory effects [5]. Higher levels of intestinal alcohol-producing bacteria promoted by alterations in the microbiota are also associated with inflammatory changes in NASH [105] and result in an increased presence of ROS in the liver leading to inflammation [20]. Under conditions of endotoxemia, TLR4 receptors can recognize LPS as a hazard signal by inducing activation of genes associated with inflammation. TLR2 and TLR9 are also important factors in the pathogenesis of human NASH [78]. On the other hand, Kupffer cells are stimulated by LPS via TLRs which can recognize PAMPs, and damage-associated molecular patterns (DAMPs). Increased levels of PAMP and DAMP (caused by disruption of the intestinal barrier and endotoxins circulating through the gut-liver axis) and their binding to TLRs, may initiate the release of pro-inflammatory cytokines (TNF-α, IL-8, IL-1β), stimulating lipid accumulation and cell death in hepatocytes resulting in NAFLD, NASH, and cirrhosis [92].

NASH Phenotype Related to HFD-Driven Microbiota Alterations

Germ-free mice present an exacerbated NASH phenotype when they are subjected to HFD and inoculated with the microbiota of NASH patients [106]. Xie and collaborators found a significantly altered gut microbiota in the progression of liver disease, showing an increase in *Firmicutes* and *Actinobacteria*, and a reduction in *Bacteroidetes* and *Proteobacteria* at the phylum level [107]. Concerning genus, *Ruminococcaceae* UCG-010 of the family *Ruminococcaceae*, order *Clostridiales*, and class *Clostridia* were found in lower concentrations in NASH patients in an Asian cohort compared to healthy controls [71]. Furthermore, several results support the hypothesis that genetics could also influence the progression of NASH. A study carried out by Qian and collaborators revealed two different liver phenotypes and several upregulated inflammation-related genes as well as altered lipid profiles in the NASH group. In addition, some lipid-related pathways were different between the NASH and NAFLD groups. *Bacteroidetes* abundance and Bacteroides genus were reduced in NASH with increased *Mucispirillum schaedleri* species [108].

It has been found that the *F11r* gene participates in the maintenance of intestinal epithelial permeability. Accordingly, an increase in inflammatory microbial taxa and liver injury was observed in *F11r* (-/-) mice triggered by HFD, fructose, and cholesterol. In mice subjected to HFD, Duparc and collaborators studied the effect of the hepatocyte-specific deletion of *MyD88*, which led to glucose intolerance, inflammation and hepatic insulin resistance, as well as changes of specific gut microbes such as a reduction in *Sutterella* and *Allobaculum* as well as an increment in *Ruminococcus* and

Oscillospira compared with normal chow diet animals [109]. In the same line, the deletion of the *p62/Sqstm1* and *Nrf2* genes was associated with the development of NASH in mice fed with a normal chow diet through the proliferation of Gram-negative bacteria and LPS in feces along with increased intestinal permeability [110].

HEPATOCELLULAR CARCINOMA

Pathophysiology

Liver cancers, among which HCC is the most frequent, are the fourth most common cause of cancer-related death worldwide. The incidence of liver cancer has increased in Western countries over the last two decades, and it is predicted to continue rising in the future [111]. HCC usually arises in patients with underlying cirrhosis. Therefore, its risk factors are in accordance with sustained liver injury risk factors, i.e., hepatitis B and C, excessive alcohol intake resulting in ALD, lifestyle patterns and physiopathological processes leading to NAFLD (as mentioned previously) and exposure to aflatoxin B1-contaminated food [112].

Notably, dysbiosis is present in patients with HCC of different etiologies [113-115]. In addition, Zheng and collaborators found that dysbiosis was more common in cirrhosis-induced HCC than in non-cirrhosis-induced HCC patients [116].

In hepatocarcinogenesis, dysbiosis not only promotes chronic disease but also progression to HCC [117-120]. Remarkably, to date, evidence for the role of the intestinal microbiota in HCC progression has been obtained principally from animal studies [121], mainly performed in rodent models of NASH-associated HCC to mimic a frequent pathogenic pathway in patients [122]. Other pathogenic HCC pathways also seem to involve dysbiosis. For example, aflatoxin, as well as other mycotoxins, can provoke gut dysbiosis and gut barrier disruption, among other biological effects [123]. Additionally, observational studies in humans and animals have linked specific dietary compounds and dietary patterns with the risk of liver cancer development, by mechanisms including dysbiosis [124]. Interestingly, a study in mice suggested that maternal feeding with a Westernized diet associated microbes during pregnancy is associated with microbes that increase the risk of HCC in the progeny, pointing to a transgenerational transmission of cancer risk that can be modulated by the gut microbiota [125].

Other mechanisms involved in the pathogenesis of HCC through the gut microbiota are mediated by IL-25. This protein, which is overexpressed in the serum and liver of HCC patients and predicts poor prognosis, could also be involved in HCC promotion through dysbiosis. On one hand, *in vitro* experiments showed that IL-25 induces HCC cell migration and invasion by facilitating M2 macrophage polarization. Besides that, antibiotic-induced dysbacteriosis activated IL-25 production by tuft cells [115]. The relationship of other immune cell types with dysbiosis and HCC has also been studied. Circulating and tumor-infiltrating CD8+ T cells from a subset of HCC patients caused by chronic hepatitis B virus infection showed response towards certain gut bacteria such as *B. longum* and *Enterococcus hirae*, implying that these bacterial species could be used as cancer immunotherapy due to their capacity to stimulate antitumor responses [126].

Importantly, sex-dependent differences in gut microbiota have been observed. Indeed, intestinal microbiota, liver transcriptome and HCC growth differ according to sex and dietary fat type (polyunsaturated or saturated fat) in diethylnitrosamine and HFD-treated mice. A high saturated fat diet was associated with higher liver tumorigenesis only in males [127]. Moreover, changes in the microbiota were different in male and female liver-specific *Tsc1* knockout mice (these mice spontaneously develop HCC at 9-10 months of age) [128]. Similarly, differences in the gut microbiota involved in bile acid metabolism were observed between non-treated male and female mice and were accentuated in the streptozotocin-HFD-induced NASH-HCC model [129]. Accordingly, these results could explain, at least partially, the disparity in HCC development in male and female individuals.

Alteration in Microbiota and Its Effects on Progression to HCC

It has been postulated that *E. coli* might promote hepatocarcinogenesis, since HCC patients with underlying cirrhosis present overgrowth of *E. coli* when compared with cirrhotic patients [130]. The same comparison revealed a microbiome pattern linked to inflammation in HCC patients [131]. Other descriptive studies have examined the differences in microbiota composition between HCC, cirrhotic and healthy individuals, but further mechanistic studies are needed to establish a causative effect of specific microbial species in hepatocarcinogenesis [132]. Similar studies have been performed in animal models; for instance, a streptozotocin-HFD-induced NASH-HCC mouse model [107]. *Bacteroides* genus may also play a role in HCC development, since the Shaoyao Ruangan mixture, an anticancer therapy, decreased the abundance of *Bacteroides* in an HCC murine model induced by overexpression of a mutant *Ras* oncogene in hepatocytes [133]. Finally, data from a transgenic murine model of hepatitis B virus-associated HCC showed that HCC was aggravated by oral administration of certain intestinal bacteria (i.e., *Helicobacter hepaticus*), mainly via activation of cytokine production in intrahepatic natural killer cells, pointing to a role of dysbiosis in perpetuating a vicious circle of hepatocarcinogenesis [134].

Role of Bile Acids in HCC

Mounting evidence shows that the secondary bile acid ursodeoxycholic acid, produced by the transformation of (cheno)deoxycholic acid by gut bacteria, regulates intestinal barrier integrity and lipid metabolism. Accordingly, ursodeoxycholic acid has been proposed as a treatment for HCC [135].

On the other hand, FXR knockout mice presented macroscopically visible liver tumors after 15 months of Western diet feeding. Both Western diet intake and FXR deficiency caused hepatitis, gut dysbiosis and reduced butyrate production, pointing to a role of dysbiosis-associated dysregulated bile acid synthesis in hepatic inflammation promotion, which, in turn, contributes to carcinogenesis [136]. Indeed, intrahepatic retention of a myriad of hydrophobic bile acids was enhanced in a streptozotocin and HFD-induced NASH-HCC murine model, while increasing intestinal excretion of these hydrophobic bile acids suppressed HCC formation [107].

Previous studies have also shown that obesity-induced dysbiosis increases deoxycholic acid, a microbiota-produced secondary bile acid that causes DNA damage, which then translocates to the liver and provokes the senescence-associated secretory phenotype (SASP) in hepatic stellate cells. Senescent HSCs secrete pro-inflammatory factors and create a tumor-promoting microenvironment, facilitating HCC in mice exposed to the chemical carcinogen 7,12-dimethylbenz(a)anthracene [137]. Loo and collaborators found that the Gram-positive gut microbial component lipoteichoic acid is also induced in obesity. In this study, lipoteichoicacid promoted HCC development by translocating to the liver and enhancing the SASP of HSCs. In collaboration with the obesity-induced gut bacterial metabolite deoxycholic acid, lipoteichoic acid upregulated the expression of SASP factors and COX2 via TLR2. Activation of the COX2 pathway produced prostaglandin E_2 (PGE$_2$) in senescent HSCs in the tumor microenvironment, thereby inhibiting antitumor immunity and generating a pro-tumorigenic microenvironment [138].

Supplemental Approaches to Suppress Hepatocarcinogenesis by Modulation of the Gut Microbiota

Zinc(II)

Oral administration of zinc(II) complexes of curcumin have antitumoral effects in the HepG2 liver xenograft tumor mouse model. These effects were abolished after gut microbiome depletion with antibiotics. Furthermore, zinc(II) complexes of curcumin ameliorated dysbiosis and zinc dyshomeostasis in a rat HCC model, pointing to modulation of microbiota-mediated zinc homeostasis as a possible mechanism underlying the antitumoral effects of zinc(II) complexes of curcumin [139].

Superantigen-Like Protein 6

Regarding therapeutic effects, the microbiota-derived *Staphylococcal* superantigen-like protein 6 presented antitumoral effects in combination with sorafenib in HCC-bearing mice, probably via downregulation of PI3K/Akt-mediated glycolysis by blocking CD47, as observed *in vitro* in HCC cell lines [140].

Probiotics

Similar to chronic liver diseases, probiotics have also been proposed as a treatment for HCC. For instance, the probiotic mixture Prohep shifted the gut microbial community towards specific beneficial bacteria and impaired tumor growth by reducing the T helper 17 (Th17) cell population in the gut and downregulating IL-17 cytokine production [141].

Butyrate

In vitro, the microbiota product butyrate induced autophagy of the HCC cell line Huh7 by increasing intracellular ROS levels [142]. These ROS generated by butyrate administration can induce apoptosis by activating miR-22 expression [143], suggesting that butyrate could not only be a natural barrier for tumorigenesis but also an effective treatment for HCC, as well as for other chronic liver diseases.

Dietary Fibers

Dietary soluble fibers are fermented by intestinal bacteria into SCFAs, and therefore, are considered to have positive effects on health status. Nevertheless, an inulin-enriched high-fat diet caused dysbiosis and HCC in mice. Inhibition of fermentation by pharmacological treatment or depletion of fermenting bacteria decreased intestinal SCFAs and protected against HCC. These results have led to controversy regarding the benefits of the enrichment of foods with fermentable fiber [144].

Effects of Microbiota Depletion by Antibiotics

Microbiota depletion by antibiotic treatment improved liver injury and decreased tumor foci number in an HCC mouse model generated by feeding mice with a novel diet, the steatohepatitis-inducing HFD (named STHD-01) [145], for 41 weeks, probably by attenuating STHD-01-induced oxidative stress [146]. Antibiotics also reduced the accumulation of secondary bile acids, as well as secondary bile acids-induced mTOR pathway activation in hepatocytes, both caused by STHD-01 feeding [147]. The antitumoral effect of antibiotics has been verified in other NASH-driven HCC murine models bearing dysbiosis [148], as well as in a HCC model of inulin-fed TLR5-deficient mice [149]. In contrast, it has been hypothesized that antibiotic treatment may change microbiota composition and reduce bacterial diversity in humans, leading to dysbiosis and hepatocarcinogenesis [150]. Nonetheless, no association was found between antibiotic use and the risk of liver cancer in a study in humans [151]. In line with the idea that microbiota depletion inhibits HCC development, hepatocarcinogenesis was abolished in liver-specific *Trim28* (–/–) mice subjected to HFD raised in germ-free conditions, compared to conventionally-raised counterparts that developed tumors with more than 50% penetrance [152]. Ma and collaborators confirmed that depletion of commensal gut bacteria by an antibiotic cocktail suppressed tumor growth in the liver. In a MYC transgenic mouse model of spontaneous HCC, antibiotics induced a selective increase of hepatic CXCR6-positive natural killer T (NKT) cells. On antigen stimulation, these NKT cells with an activated phenotype and increased production of interferon-γ (IFN-γ) were responsible for the antitumoral effect of the antibiotic cocktail. Since intestinal bacteria can transform primary into secondary bile acids, the authors proposed that microbiota-produced bile acids serve as messengers to control the chemokine-dependent increase of NKT cells in the liver and, therefore, confer immunity against liver cancer. This mechanism may apply not only to primary liver tumors but also to liver metastasis [153]. In contrast, since the microbiota participates in arsenic metabolism, depletion of the microbiota with antibiotics alters arsenic biotransformation and increases its toxic effects, which may be linked to an augmented risk of arsenic-induced HCC in mice [154].

TREATMENT APPROACHES

Despite the numerous studies mentioned above, the role of microbiota in the pathophysiology of liver disease remains controversial since the mechanisms and beneficial effects that the gut microbial population can exert on their treatment and progression are still unknown. Exercise and caloric restriction have currently been proposed focused on body weight loss to improve steatosis status [78] and seem to be the most successful care recommendations. In addition, bariatric surgery induces weight loss and a better metabolic profile. There is a growing interest in understanding the modifications that this technique produces in intestinal microbiota with promising results [155, 156].

The concept of disease transmission by FMT, together with its potential and promising therapeutic role, is also being explored [157]. The use of FMT has recently come into the spotlight and could be an important key in preventing the progression of NAFLD [158, 159]. For instance, a study by Zhou and collaborators performed in HFD mice revealed a reduction of lipid intracellular accumulation and pro-inflammatory cytokines as well as corrected gut microbiota disturbance after FMT [30]. In contrast, to date, human FMT presents several limitations such as the lack of consensus in stool donor preparation and transplantation method (endoscopy, nasointestinal tubes, or capsules). Besides, the potential long-term adverse effects need to be studied in depth. Figure 1 summarizes the current strategies reported to revert dysbiosis in humans.

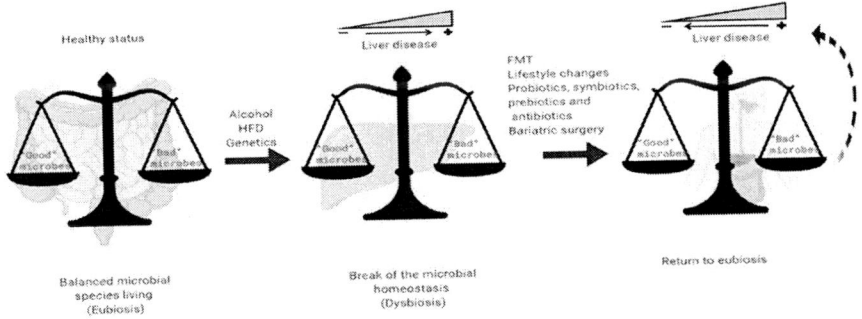

Figure 1. Schematic representation showing the current treatments to restore healthy gut microbiota in liver disease.

CONCLUSION

Liver diseases affect several million people worldwide but their exact cause has yet to be elucidated and it is almost certainly not the same in every patient. Although hopeful results have been achieved in the understanding of the pathophysiology and treatment of liver diseases, in many cases researchers do not take into account the limitations of biomarkers used for the diagnosis and progression of NAFLD. The huge variability and heterogeneity present in these studies also hamper the homogeneity in results.

It has been described that the gut microbiota plays a role in the pathogenesis of liver diseases, reporting gut dysbiosis as an important inductor of these diseases. However, independently of its etiology (i.e., alcohol abuse or fat liver accumulation), chronic liver disease may progress to more severe disorders, such as NASH, progressive fibrosis, cirrhosis, and hepatic cancer [27]. Indeed, evidence indicates that the onset, development, and severity, as well as the incidence rate of liver diseases, are closely associated with the gut microbiota. Although current strategies or nutraceutical supplementations for ALD and NAFLD have yielded promising results (i.e., FMT; probiotics, symbiotic, prebiotic and antibiotic intake; changes in lifestyle, etc.), the therapeutic perspectives for patients require a general view of the liver disease spectrum. It is also necessary to define the pathogenic mechanisms related to alterations in the gut microbiota that are responsible for disease progression and severity. Only under this premise, will we be able to identify patients with the risk of progressing to end-stage liver disease and act accordingly.

To date, metagenomics (or 16S-targeted sequencing) and metabolomics are useful tools to decipher the role of the bacterial species involved in liver functionality. The gut microbes identified and their metabolites could serve as biomarkers to assist in disease diagnostics as well as in the design of approaches to restore healthy microbiota. Moreover, this knowledge could be helpful to prevent the progression of liver disease and selectively avoid the impairment of liver-related damage such as oxidative stress, inflammation, fibrosis, and apoptosis.

To conclude, more in depth research is needed to provide practical strategies to manage liver diseases with modulation of the gut microbiota. These data could provide an opportunity to develop comprehensive and consistent identification of the specific factors supporting the onset and progression of liver diseases and serve as therapeutic targets for their treatment.

REFERENCES

[1]   Tripathi, Anupriya, Justine Debelius, David A. Brenner, Michael Karin, Rohit Loomba, Bernd Schnabl, and Rob Knight. 2018. "The gut-liver axis and the intersection with the microbiome." *Nature Reviews Gastroenterology and Hepatology* volume 15(7):397-411. doi: 10.1038/s41575-018-0011-z.

[2]   Delzenne, Nathalie M., Christelle Knudsen, Martin Beaumont, Julie Rodriguez, Audrey M. Neyrinck, and Laure B. Bindels. 2019. "Contribution of the gut microbiota to the regulation of host metabolism and energy balance: A focus on the gut-liver axis." *Proceedings of the Nutrition Society* 78(3):319-28. doi: 10.1017/S0029665118002756.

[3] Baffy, Gyorgy. 2019. "Potential mechanisms linking gut microbiota and portal hypertension." *Liver International* 39(4):598-609. doi: 10.1111/liv.13986.

[4] Albillos, Agustín, Andrea de Gottardi, María Rescigno. 2020. "The gut-liver axis in liver disease: Pathophysiological basis for therapy." *Journal of Hepatology* 72(3):558-77. doi: 10.1016/j.jhep.2019.10. 003.

[5] Duarte, Sebastião M. B., Jose Tadeu Stefano, and Claudia P. Oliveira. 2019. "Microbiota and nonalcoholic fatty liver disease/nonalcoholic steatohepatitis (NAFLD/NASH)." *Annals of Hepatology* 18(3):416-21. doi: 10.1016/j.aohep.2019.04.006.

[6] Hu, Haiming, Aizhen Lin, Mingwang Kong, Xiaowei Yao, Mingzhu Yin, Hui Xia, Jun Ma, and Hongtao Liu. 2020. "Intestinal microbiome and NAFLD: molecular insights and therapeutic perspectives." *Journal of Gastroenterology* 55(2):142-58. doi: 10.1007/s00535-019-01649-8.

[7] Wang, Baohong, Mingfei Yao, Longxian Lv, Zongxin Ling, and Lanjuan Li. 2017. "The Human Microbiota in Health and Disease." *Engineering* 3(1):71-82. doi: 10.1016/J.ENG.2017.01.008.

[8] Cornide-Petronio, María Eugenia, Ana Isabel Álvarez-Mercado, Mónica B. Jiménez-Castro, and Carmen Peralta. 2020. "Current Knowledge about the Effect of Nutritional Status, Supplemented Nutrition Diet, and Gut Microbiota on Hepatic Ischemia-Reperfusion and Regeneration in Liver Surgery." *Nutrients* 12(2): 284. doi: 10. 3390/nu12020284.

[9] Bluemel, Sena, Brandon Williams, Rob Knight, and Bernd Schnabl. 2016. "Precision medicine in alcoholic and nonalcoholic fatty liver disease via modulating the gut microbiota." *American Journal of Physiology - Gastrointestinal and Liver Physiology* 311(6):G1018–36. doi: 10.1152/ajpgi.00245.2016.

[10] Álvarez-Mercado, Ana Isabel, Miguel Navarro-Oliveros, Cándido Robles-Sánchez, Julio Plaza-Díaz, María José Sáez-Lara, Sergio Muñoz-Quezada, Luis Fontana, and Francisco Abadía-Molina. 2019. "Microbial Population Changes and Their Relationship with Human

Health and Disease." *Microorganisms* 7(3):68. doi: 10.3390/microorganisms7030068.

[11] Meroni, Marica, Miriam Longo, and Paola Dongiovanni. 2019. "Alcohol or Gut Microbiota: Who Is the Guilty?" *International Journal of Molecular Sciences* 20(18):E4568. doi: 10.3390/ijms 20184568.

[12] Adolph, Timon E., Christoph Grander, Alexander R. Moschen, and Herbert Tilg. 2018. "Liver–Microbiome Axis in Health and Disease." *Trends in Immunology* 39(9):712-23. doi: 10.1016/j.it. 2018.05.002.

[13] Stärkel, Peter, Sophie Leclercq, Philippe de Timary, and Bernd Schnabl. 2018. "Intestinal Dysbiosis and Permeability: The Yin and Yang in Alcohol Dependence and Alcoholic Liver Disease." *Clinical Science* 132(2):199–212. doi: 10.1042/CS20171055.

[14] Hartmann, Phillipp, Caroline T. Seebauer, and Bernd Schnabl. 2015. "Alcoholic Liver Disease: The Gut Microbiome and Liver Cross Talk." *Alcoholism: Clinical and Experimental Research* 39(5):763–75. doi: 10.1111/acer.12704.

[15] Dubinkina, Veronika B., Alexander V. Tyakht, Vera Y. Odintsova, Konstantin S. Yarygin, Boris A. Kovarsky, Alexander V. Pavlenko, Dmitry S. Ischenko, et al. 2017. "Links of Gut Microbiota Composition with Alcohol Dependence Syndrome and Alcoholic Liver Disease." *Microbiome* 5(1):141. doi: 10.1186/s40168-017-0359-2.

[16] Barr, Tasha, Suhas Sureshchandra, Paul Ruegger, Jingfei Zhang, Wenxiu Ma, James Borneman, Kathleen Grant, and Ilhem Messaoudi. 2018. "Concurrent Gut Transcriptome and Microbiota Profiling Following Chronic Ethanol Consumption in Nonhuman Primates." *Gut Microbes* 9(4):338-56. doi: 10.1080/19490976.2018. 1441663.

[17] Bajaj, Jasmohan S. 2019. "Alcohol, Liver Disease and the Gut Microbiota." *Nature Reviews Gastroenterology and Hepatology* 16(4):235–46. doi: 10.1038/s41575-018-0099-1.

[18] Litwinowicz, Kamil, Marcin Choroszy, and Ewa Waszczuk. 2020. "Changes in the Composition of the Human Intestinal Microbiome in Alcohol Use Disorder: A Systematic Review." *American Journal of*

Drug and Alcohol Abuse 46(1):4–12. doi: 10.1080/00952990.2019. 1669629.

[19] Betrapally, Naga S., Patrick M. Gillevet, and Jasmohan S. Bajaj. 2016. "Changes in the Intestinal Microbiome and Alcoholic and Nonalcoholic Liver Diseases: Causes or Effects?" *Gastroenterology* 150(8):1745-55.e3. doi: 10.1053/j.gastro.2016.02.073.

[20] Fukui, Hiroshi. 2019. "Role of Gut Dysbiosis in Liver Diseases: What Have We Learned So Far?" *Diseases* 7(4):E58. doi: 10.3390/ diseases7040058.

[21] Guohong-Liu, Qingxi-Zhao, and Hongyun-Wei. 2019. "Characteristics of Intestinal Bacteria with Fatty Liver Diseases and Cirrhosis." *Annals of Hepatology* 18(6):796–803. doi: 10.1016/j. aohep.2019.06.020.

[22] Albhaisi, Somaya A.M., Jasmohan S. Bajaj, and Arun J. Sanyal. 2020. "Role of Gut Microbiota in Liver Disease." *American Journal of Physiology. Gastrointestinal and Liver Physiology* 318(1):G84–98. doi: 10.1152/ajpgi.00118.2019.

[23] Dabur, Rajesh, Amey Shirolkar, Vijender Mishra, and Baljeet Singh Yadav. 2018. "Non-Invasive Qualitative Urinary Metabolomic Profiling Discriminates Gut Microbiota Derived Metabolites in the Moderate and Chronic Alcoholic Cohorts." *Current Pharmaceutical Biotechnology* 18(14):1175–89. doi: 10.2174/1389201019666180308 093207.

[24] Deda, Olga, Christina Virgiliou, Amvrosios Orfanidis, and Helen G. Gika. 2019. "Study of Fecal and Urinary Metabolite Perturbations Induced by Chronic Ethanol Treatment in Mice by UHPLC-MS/MS Targeted Profiling." *Metabolites* 9(10):E232. doi: 10.3390/ metabo9100232.

[25] Zhang, Xiao, Koji Yasuda, Robert A Gilmore, Susan V Westmoreland, Donna M Platt, Gregory M Miller, and Eric J Vallender. 2019. "Alcohol-Induced Changes in the Gut Microbiome and Metabolome of Rhesus Macaques." *Psychopharmacology* 236(5):1531–44. doi: 10.1007/s00213-019-05217-z.

[26]   Aron-Wisnewsky, Judith, Moritz V. Warmbrunn, Max Nieuwdorp, Karine Clément. 2020. "Nonalcoholic Fatty Liver Disease: Modulating Gut Microbiota to Improve Severity?" *Gastroenterology* 158(7):1881–98. doi: 10.1053/j.gastro.2020.01.049.

[27]   Mathurin, Philippe, and Ramon Bataller. 2015. "Trends in the Management and Burden of Alcoholic Liver Disease." *Journal of Hepatology* 62(1 Suppl):S38–46. doi: 10.1016/j.jhep.2015.03.006.

[28]   Chen, Peng, Yukiko Miyamoto, Magdalena Mazagova, Kuei Chuan Lee, Lars Eckmann, and Bernd Schnabl. 2015. "Microbiota Protects Mice against Acute Alcohol-Induced Liver Injury." *Alcoholism: Clinical and Experimental Research* 39(12):2313–23. doi: 10.1111/ acer.12900.

[29]   Bala, Shashi, Miguel Marcos, Arijeet Gattu, Donna Catalano, and Gyongyi Szabo. 2014. "Acute Binge Drinking Increases Serum Endotoxin and Bacterial DNA Levels in Healthy Individuals." *PLoS One* 9(5):e96864. doi: 10.1371/journal.pone.0096864.

[30]   Zhou, Zhanxiang, and Wei Zhong. 2017. "Targeting the Gut Barrier for the Treatment of Alcoholic Liver Disease." *Liver Research* 1(4):197-207. doi: 10.1016/j.livres.2017.12.004.

[31]   Lang, Sonja, Yi Duan, Jinyuan Liu, Manolito G. Torralba, Claire Kuelbs, Meritxell Ventura-Cots, Juan G. Abraldes, et al. 2020. "Intestinal Fungal Dysbiosis and Systemic Immune Response to Fungi in Patients With Alcoholic Hepatitis." *Hepatology* 71(2):522– 38. doi: 10.1002/hep.30832.

[32]   Chu, Huikuan, Yi Duan, Sonja Lang, Lu Jiang, Yanhan Wang, Cristina Llorente, Jinyuan Liu, et al. 2020. "The Candida Albicans Exotoxin Candidalysin Promotes Alcohol-Associated Liver Disease." *Journal of Hepatology* 72(3):391–400. doi: 10.1016/j.jhep.2019. 09.029.

[33]   Ciocan, Dragos, Cosmin Sebastian Voican, Laura Wrzosek, Cindy Hugot, Dominique Rainteau, Lydie Humbert, Anne Marie Cassard, and Gabriel Perlemuter. 2018. "Bile Acid Homeostasis and Intestinal Dysbiosis in Alcoholic Hepatitis." *Alimentary Pharmacology and Therapeutics* 48(9):961–74. doi: 10.1111/apt. 14949.

[34] Guo, Feifei, Kang Zheng, Raquel Benedé-Ubieto, Francisco Javier Cubero, and Yulia A. Nevzorova. 2018. "The Lieber-DeCarli Diet— A Flagship Model for Experimental Alcoholic Liver Disease." *Alcoholism: Clinical and Experimental Research* 42(10):1828–40. doi: 10.1111/acer.13840.

[35] Bertola, Adeline, Stephanie Mathews, Sung Hwan Ki, Hua Wang, and Bin Gao. 2013. "Mouse Model of Chronic and Binge Ethanol Feeding (the NIAAA Model)." *Nature Protocols* 8(3):627–37. doi: 10.1038/nprot.2013.032.

[36] Tsukamoto, Hidekazu, Samuel W. French, Roger D. Rektelberger, and Corey Largman. 1985. "Cyclical Pattern of Blood Alcohol Levels during Continuous Intragastric Ethanol Infusion in Rats." *Alcoholism: Clinical and Experimental Research* 9(1):31-7. doi: 10.1111/j.1530-0277.1985.tb05046.x.

[37] Becker, Howard C., and Dorit Ron. 2014. "Animal Models of Excessive Alcohol Consumption: Recent Advances and Future Challenges." *Alcohol* 48(3):205–8. doi: 10.1016/j.alcohol.2014.04.001.

[38] Gao, Bin, Ming Jiang Xu, Adeline Bertola, Hua Wang, Zhou Zhou, and Suthat Liangpunsakul. 2017. "Animal Models of Alcoholic Liver Disease: Pathogenesis and Clinical Relevance." *Gene Expression* 17(3):173–86. doi: 10.3727/105221617X695519.

[39] Lamas-Paz, Arantza, Fengjie Hao, Leonard J. Nelson, Maria Teresa Vázquez, Santiago Canals, Manuel Gómez del Moral, Eduardo Martínez-Naves, Yulia A. Nevzorova, and Francisco Javier Cubero. 2018. "Alcoholic Liver Disease: Utility of Animal Models." *World Journal of Gastroenterology* 24(45):5063–75. doi: 10.3748/wjg.v24.i45.5063.

[40] Posteraro, Brunella, Francesco Paroni Sterbini, Valentina Petito, Stefano Rocca, Tiziana Cubeddu, Cristina Graziani, Vincenzo Arena, et al. 2018. "Liver Injury, Endotoxemia, and Their Relationship to Intestinal Microbiota Composition in Alcohol-Preferring Rats." *Alcoholism: Clinical and Experimental Research* 42(12):2313–25. doi: 10.1111/acer.13900.

[41] Thiele, Todd E., and Montserrat Navarro. 2014. "'Drinking in the Dark' (DID) Procedures: A Model of Binge-like Ethanol Drinking in Non-Dependent Mice." *Alcohol* 48(3):235–41. doi: 10.1016/j. alcohol.2013.08.005.

[42] Summa, Keith C., Robin M. Voigt, Christopher B. Forsyth, Maliha Shaikh, Kate Cavanaugh, Yueming Tang, Martha Hotz Vitaterna, Shiwen Song, Fred W. Turek, and Ali Keshavarzian. 2013. "Disruption of the Circadian Clock in Mice Increases Intestinal Permeability and Promotes Alcohol-Induced Hepatic Pathology and Inflammation." *PLoS ONE* 8(6):e67102. doi: 10.1371/journal.pone. 0067102.

[43] Swanson, Garth, Christopher B. Forsyth, Yueming Tang, Maliha Shaikh, Lijuan Zhang, Fred W. Turek, and Ali Keshavarzian. 2011. "Role of Intestinal Circadian Genes in Alcohol-Induced Gut Leakiness." *Alcoholism: Clinical and Experimental Research* 35(7):1305–14. doi: 10.1111/j.1530-0277.2011.01466.x.

[44] Stadlbauer, Vanessa, Angela Horvath, Irina Komarova, Bianca Schmerboeck, Nicole Feldbacher, Sonja Wurm, Ingeborg Klymiuk, et al. 2019. "A Single Alcohol Binge Impacts on Neutrophil Function without Changes in Gut Barrier Function and Gut Microbiome Composition in Healthy Volunteers." *PLoS ONE* 14(2):e0211703. doi: 10.1371/journal.pone.0211703.

[45] Lee, Jang Eun, Jung Su Ha, Ho Young Park, and Eunjung Lee. 2020. "Alteration of Gut Microbiota Composition by Short-Term Low-Dose Alcohol Intake Is Restored by Fermented Rice Liquor in Mice." *Food Research International (Ottawa, Ont.)* 128:108800. doi: 10. 1016/j.foodres.2019.108800.

[46] Samuelson, Derrick R., Min Gu, Judd E. Shellito, Patricia E. Molina, Christopher M. Taylor, Meng Luo, and David A. Welsh. 2019. "Intestinal Microbial Products From Alcohol-Fed Mice Contribute to Intestinal Permeability and Peripheral Immune Activation." *Alcoholism: Clinical and Experimental Research* 43(10):2122–33. doi: 10.1111/acer.14176.

[47] Inamine, Tatsuo, An Ming Yang, Lirui Wang, Kuei Chuan Lee, Cristina Llorente, and Bernd Schnabl. 2016. "Genetic Loss of Immunoglobulin A Does Not Influence Development of Alcoholic Steatohepatitis in Mice." *Alcoholism: Clinical and Experimental Research* 40(12):2604–13. doi: 10.1111/acer.13239.

[48] Su, Qian qian, Yang yang Tian, Zhen ni Liu, Lei lei Ci, and Xiong wen Lv. 2019. "Purinergic P2X7 Receptor Blockade Mitigates Alcohol-Induced Steatohepatitis and Intestinal Injury by Regulating MEK1/2-ERK1/2 Signaling and Egr-1 Activity." *International Immunopharmacology* 66:52–61. doi: 10.1016/j.intimp.2018.11.012.

[49] Yang, Xiaoli, Fang He, Yanting Zhang, Jing Xue, Ke Li, Xiaoxia Zhang, Lili Zhu, Zhen Wang, Hao Wang, and Shaoqi Yang. 2019. "Inulin Ameliorates Alcoholic Liver Disease via Suppressing LPS-TLR4-Mψ Axis and Modulating Gut Microbiota in Mice." *Alcoholism: Clinical and Experimental Research* 43(3):411–24. doi: 10.1111/acer.13950.

[50] Wang, Lirui, Derrick E. Fouts, Peter Stärkel, Phillipp Hartmann, Peng Chen, Cristina Llorente, Jessica DePew, et al. 2016. "Intestinal REG3 Lectins Protect against Alcoholic Steatohepatitis by Reducing Mucosa-Associated Microbiota and Preventing Bacterial Translocation." *Cell Host and Microbe* 19(2):227–39. doi: 10.1016/j.chom.2016.01.003.

[51] Zhong, Wei, Xiaoyuan Wei, Liuyi Hao, Tai Du Lin, Ruichao Yue, Xinguo Sun, Wei Guo, et al. 2019. "Paneth Cell Dysfunction Mediates Alcohol-Related Steatohepatitis Through Promoting Bacterial Translocation in Mice: Role of Zinc Deficiency." *Hepatology* 71(5):1575-91. doi: 10.1002/hep.30945.

[52] Duan, Yi, Cristina Llorente, Sonja Lang, Katharina Brandl, Huikuan Chu, Lu Jiang, Richard C. White, et al. 2019. "Bacteriophage targeting of gut bacterium attenuates alcoholic liver disease." *Nature* 575(7783):505-11. doi: 10.1038/s41586-019-1742-x.

[53] Glueck, Bryan, Yingchun Han, and Gail A.M. Cresci. 2018. "Tributyrin Supplementation Protects Immune Responses and Vasculature and Reduces Oxidative Stress in the Proximal Colon of

Mice Exposed to Chronic-Binge Ethanol Feeding." *Journal of Immunology Research* 2018:9671919. doi: 10.1155/2018/9671919.

[54]   Neyrinck, Audrey M., Usune Etxeberria, Bernard Taminiau, Georges Daube, Matthias Van Hul, Amandine Everard, Patrice D. Cani, Laure B. Bindels, and Nathalie M. Delzenne. 2017. "Rhubarb Extract Prevents Hepatic Inflammation Induced by Acute Alcohol Intake, an Effect Related to the Modulation of the Gut Microbiota." *Molecular Nutrition and Food Research* 61(1). doi: 10.1002/mnfr.201500899.

[55]   Chen, Jing, Yan Han Xuan, Ming Xiao Luo, Xiang Gui Ni, Li Qian Ling, Shi Jia Hu, Jing Qiao Chen, et al. 2020. "Kaempferol Alleviates Acute Alcoholic Liver Injury in Mice by Regulating Intestinal Tight Junction Proteins and Butyrate Receptors and Transporters." *Toxicology* 429:152338. doi: 10.1016/j.tox.2019.152338.

[56]   Park, Sunmin, Da S. Kim, Xuangao Wu, and Qiu J Yi. 2018. "Mulberry and Dandelion Water Extracts Prevent Alcohol-Induced Steatosis with Alleviating Gut Microbiome Dysbiosis." *Experimental Biology and Medicine* 243(11):882–94. doi: 10.1177/1535370218789068.

[57]   Liu, Xiaoxia, Ke Zhao, Xingbin Yang, and Yan Zhao. 2019. "Gut Microbiota and Metabolome Response of Decaisnea Insignis Seed Oil on Metabolism Disorder Induced by Excess Alcohol Consumption." *Journal of Agricultural and Food Chemistry* 67(38):10667–77. doi: 10.1021/acs.jafc.9b04792.

[58]   Wang, Yuchuan, Min Guan, Xin Zhao, and Xinli Li. 2018. "Effects of Garlic Polysaccharide on Alcoholic Liver Fibrosis and Intestinal Microflora in Mice." *Pharmaceutical Biology* 56(1):325–32. doi: 10. 1080/13880209.2018.1479868.

[59]   Liu, Huilin, Meihong Liu, Xueqi Fu, Ziqi Zhang, Lingyu Zhu, Xin Zheng, and Jingsheng Liu. 2018. "Astaxanthin Prevents Alcoholic Fatty Liver Disease by Modulating Mouse Gut Microbiota." *Nutrients* 10(9):E1298. doi: 10.3390/nu10091298.

[60]   Han, Yingchun, Bryan Glueck, David Shapiro, Aaron Miller, Sanjoy Roychowdhury, and Gail A. M. Cresci. 2020. "Dietary Synbiotic Supplementation Protects Barrier Integrity of Hepatocytes and Liver

Sinusoidal Endothelium in a Mouse Model of Chronic-Binge Ethanol Exposure." *Nutrients* 12(2):E373. doi: 10.3390/nu12020373.

[61] Lowe, Patrick P., Benedek Gyongyosi, Abhishek Satishchandran, Arvin Iracheta-Vellve, Yeonhee Cho, Aditya Ambade, and Gyongyi Szabo. 2018. "Reduced Gut Microbiome Protects from Alcohol-Induced Neuroinflammation and Alters Intestinal and Brain Inflammasome Expression." *Journal of Neuroinflammation* 15(1):298. doi: 10.1186/s12974-018-1328-9.

[62] Lowe, Patrick P., Benedek Gyongyosi, Abhishek Satishchandran, Arvin Iracheta-Vellve, Aditya Ambade, Karen Kodys, Donna Catalano, Doyle V. Ward, and Gyongyi Szabo. 2017. "Alcohol-Related Changes in the Intestinal Microbiome Influence Neutrophil Infiltration, Inflammation and Steatosis in Early Alcoholic Hepatitis in Mice." *PLoS ONE* 12(3):e0174544. doi: 10.1371/journal.pone. 0174544.

[63] van den Berg, Eline H., Marzyeh Amini, Tim C. M. A. Schreuder, Robin P. F. Dullaart, Klaas Nico Faber, Behrooz Z. Alizadeh, and Hans Blokzijl. 2017. "Prevalence and determinants of non-alcoholic fatty liver disease in lifelines: A large Dutch population cohort." *PLoS ONE* 12(2):e0171502. doi: 10.1371/journal.pone.0171502.

[64] Assunção, Silvana Neves Ferraz, Ney Christian Boa Sorte, Crésio Dantas Alves, Patricia S. Almeida Mendes, Carlos Roberto Brites Alves, and Luciana Rodrigues Silva. 2017. "Nonalcoholic fatty liver disease (NAFLD) pathophysiology in obese children and adolescents: update." *Nutrición Hospitalaria [Hospital Nutrition]* 34(3):727-30. doi: 10.20960/nh.723.

[65] López-Velázquez, Jorge A., Karen V. Silva-Vidal, Guadalupe Ponciano-Rodríguez, Norberte C. Chávez-Tapia, Marco Arrese, Misael Uribe, and Nahum Méndez-Sánchez. 2014. "The prevalence of nonalcoholic fatty liver disease in the Americas." *Annals of Hepatology* 13(2):166-78. doi: 10.1016/S1665-2681(19)30879-8.

[66] Caballería, Llorenç, Guillem Pera, Maria Antònia Auladell, Pere Torán, Laura Muñoz, Dolores Miranda, Alba Alumà, et al. 2010. "Prevalence and factors associated with the presence of nonalcoholic

fatty liver disease in an adult population in Spain." *European Journal of Gastroenterology and Hepatology* 22(1):24-32. doi: 10.1097/MEG.0b013e32832fcdf0.

[67]   Pimpin, Laura, Helena Cortez-Pinto, Francesco Negro, Emily Corbould, Jeffrey V. Lazarus, Laura Webber, Nick Sheron, and EASL HEPAHEALTH Steering Committee. 2018. "Burden of liver disease in Europe: Epidemiology and analysis of risk factors to identify prevention policies." *Journal of Hepatology* 69(3):718-35. doi: 10.1016/j.jhep.2018.05.011.

[68]   Huang, Tony (Dazhong), Jason Behary, and Amany Zekry. 2019. "Non-alcoholic Fatty Liver Disease (NAFLD): A Review of Epidemiology, Risk Factors, Diagnosis and Management." *Internal Medicine Journal.* doi: 10.1111/imj.14709.

[69]   Kanda, Tatsuo, Taichiro Goto, Yosuke Hirotsu, Ryota Masuzaki, Mitsuhiko Moriyama, and Masao Omata. 2020. "Molecular Mechanisms: Connections between Nonalcoholic Fatty Liver Disease, Steatohepatitis and Hepatocellular Carcinoma." *International Journal of Molecular Sciences* 21(4):E1525. doi: 10.3390/ijms21041525.

[70]   Duarte, S. M. B., J. T. Stefano, L. Miele, F. R. Ponziani, M. Souza-Basqueira, L. S. R. R. Okada, F. G. de Barros Costa, et al. 2018. "Gut microbiome composition in lean patients with NASH is associated with liver damage independent of caloric intake: A prospective pilot study." *Nutrition, Metabolism and Cardiovascular Diseases* 28(4):369-84. doi: 10.1016/j.numecd.2017.10.014.

[71]   Tsai, Ming-Chao, Yu-Yin Liu, Chih-Che Lin, Chih-Chi Wang, Yi-Ju Wu, Chee-Chien Yong, Kuang-Den Chen, et al. 2020. "Gut Microbiota Dysbiosis in Patients with Biopsy-Proven Nonalcoholic Fatty Liver Disease: A Cross-Sectional Study in Taiwan." *Nutrients* 12(3):E820. doi: 10.3390/nu12030820.

[72]   Wang, Caihua, Chunpeng Zhu, Liming Shao, Jun Ye, Yimin Shen, and Yuezhong Ren. 2019. "Role of Bile Acids in Dysbiosis and Treatment of Nonalcoholic Fatty Liver Disease." *Mediators of Inflammation* 2019:7659509. doi: 10.1155/2019/7659509.

[73] Shama, Samaa, and Wanqing Liu. 2020. "Omega-3 Fatty Acids and Gut Microbiota: A Reciprocal Interaction in Nonalcoholic Fatty Liver Disease." *Digestive Diseases and Sciences* 65(3):906-10. doi: 10. 1007/s10620-020-06117-5.

[74] Byrne, Christopher D., and Giovanni Targher. 2015. "NAFLD: A multisystem disease." *Journal of Hepatology* 62(1 Suppl):S47-64. doi: 10.1016/j.jhep.2014.12.012.

[75] Safari, Zahra, Magali Monnoye, Peter M. Abuja, Mahendra Mariadassou, Karl Kashofer, Philippe Gérard, and Kurt Zatloukal. 2019. "Steatosis and gut microbiota dysbiosis induced by high-fat diet are reversed by 1-week chow diet administration." *Nutrition Research* 71:72-88. doi: 10.1016/j.nutres.2019.09.004.

[76] Quesada-Vázquez, Sergio, Gerard Aragonès, Josep M Del Bas, and Xavier Escoté. 2020. "Diet, Gut Microbiota and Non-Alcoholic Fatty Liver Disease: Three Parts of the Same Axis." *Cells* 9(1):E176. doi: 10.3390/cells9010176.

[77] Yuan, Jing, Chen Chen, Jinghua Cui, Jing Lu, Chao Yan, Xiao Wei, Xiangna Zhao, et al. 2019. "Fatty Liver Disease Caused by High-Alcohol-Producing Klebsiella pneumoniae." *Cell Metabolism* 30(4):675-88.e7. doi: 10.1016/j.cmet.2019.08.018.

[78] Borrelli, Antonella, Patrizia Bonelli, Franca Maria Tuccillo, Ira D. Goldfine, Joseph L. Evans, Franco Maria Buonaguro, and Aldo Mancini. 2018. "Role of gut microbiota and oxidative stress in the progression of non-alcoholic fatty liver disease to hepatocarcinoma: Current and innovative therapeutic approaches." *Redox Biology* 15:467-79. doi: 10.1016/j.redox.2018.01.009.

[79] Hoffman, Jessie B., Michael C. Petriello, Andrew J. Morris, M. Abdul Mottaleb, Yipeng Sui, Changcheng Zhou, Pan Deng, Chunyan Wang, and Bernhard Hennig. 2020. "Prebiotic inulin consumption reduces dioxin-like PCB 126-mediated hepatotoxicity and gut dysbiosis in hyperlipidemic Ldlr deficient mice." *Environmental Pollution* 261:114183. doi: 10.1016/j.envpol.2020. 114183.

[80] Porras, David, Esther Nistal, Susana Martínez-Flórez, Sandra Pisonero-Vaquero, José Luis Olcoz, Ramiro Jover, Javier González-

Gallego, María Victoria García-Mediavilla, and Sonia Sánchez-Campos. 2017. "Protective effect of quercetin on high-fat diet-induced non-alcoholic fatty liver disease in mice is mediated by modulating intestinal microbiota imbalance and related gut-liver axis activation." *Free Radical Biology and Medicine* 102:188-202. doi: 10.1016/j.freeradbiomed.2016.11.037.

[81]   Liu, Qun, Shousheng Liu, Lizhen Chen, Zhenzhen Zhao, Shuixian Du, Quanjiang Dong, Yongning Xin, and Shiying Xuan. 2019. "Role and effective therapeutic target of gut microbiota in NAFLD/NASH." *Experimental and Therapeutic Medicine* 18(3):1935-44. doi: 10. 3892/etm.2019.7781.

[82]   Chen, Yi-Hsun, Yu-Chih Wang, Chien-Chao Chiu, Yen-Peng Lee, Shao-Wen Hung, Chi-Chang Huang, Ching-Feng Chiu, Ter-Hsin Chen, Wen-Ching Huang, and Hsiao-Li Chuang. 2020. "Housing condition-associated changes in gut microbiota further affect the host response to diet-induced nonalcoholic fatty liver." *The Journal of Nutritional Biochemistry* 79:108362. doi: 10.1016/j.jnutbio.2020. 108362.

[83]   Van De Wier, Bregje, Ger H. Koek, Aalt Bast, and Guido R. M. M. Haenen. 2017. "The potential of flavonoids in the treatment of non-alcoholic fatty liver disease." *Critical Reviews in Food Science and Nutrition* 57(4):834-55. doi: 10.1080/10408398.2014.952399.

[84]   Smith, Brennan K., Katarina Marcinko, Eric M. Desjardins, James S. Lally, Rebecca J. Ford, and Gregory R. Steinberg. 2016. "Treatment of nonalcoholic fatty liver disease: role of AMPK." *American Journal of Physiology - Endocrinology and Metabolism* 311(4):E730-40. doi: 10.1152/ajpendo.00225.2016.

[85]   Yin, Xiaohan, Weiyao Liao, Qingrong Li, Hongmin Zhang, Zihui Liu, Xinjie Zheng, Lin Zheng, and Xiang Feng. 2020. "Interactions between resveratrol and gut microbiota affect the development of hepatic steatosis: A fecal microbiota transplantation study in high-fat diet mice." *Journal of Functional Foods* 67:103883. doi: 10.1016/j. jff.2020.103883.

[86] Amanullah, Iram, Yusra Habib Khan, Iqraa Anwar, Aqsa Gulzar, Tauqeer Hussain Mallhi, and Ahsan Aftab Raja. 2019. "Effect of vitamin E in non-alcoholic fatty liver disease: a systematic review and meta-analysis of randomised controlled trials." *Postgraduate Medical Journal* 95(1129):601-11. doi: 10.1136/postgradmedj-2018-136364.

[87] Jensen, Thomas, Manal F. Abdelmalek, Shelby Sullivan, Kristen J. Nadeau, Melanie Green, Carlos Roncal, Takahiko Nakagawa, et al. 2018. "Fructose and sugar: A major mediator of non-alcoholic fatty liver disease." *Journal of Hepatology* 68(5):1063-75. doi: 10.1016/j. jhep.2018.01.019.

[88] Jayakumar, Saumya, and Rohit Loomba. 2019. "Review article: emerging role of the gut microbiome in the progression of nonalcoholic fatty liver disease and potential therapeutic implications." *Alimentary Pharmacology and Therapeutics* 50(2):144-58. doi: 10.1111/apt.15314.

[89] Campo, Lyna, Sara Eiseler, Tehilla Apfel and Nikolaos Pyrsopoulos. 2019. "Fatty Liver Disease and Gut Microbiota: A Comprehensive Update." *Journal of Clinical and Translational Hepatology* 7(1):56-60. doi: 10.14218/JCTH.2018.00008.

[90] Cho, Clara E., and Marie A. Caudill. 2017. "Trimethylamine-N-Oxide: Friend, Foe, or Simply Caught in the Cross-Fire?" *Trends in Endocrinology and Metabolism* 28(2):121-30. doi: 10.1016/j.tem. 2016.10.005.

[91] Tan, Xuying, Yan Liu, Jingan Long, Si Chen, Gongcheng Liao, Shangling Wu, Chunlei Li, Lijun Wang, Wenhua Ling, and Huilian Zhu. 2019. "Trimethylamine N-Oxide Aggravates Liver Steatosis through Modulation of Bile Acid Metabolism and Inhibition of Farnesoid X Receptor Signaling in Nonalcoholic Fatty Liver Disease." *Molecular Nutrition and Food Research* 63(17):e1900257. doi: 10.1002/mnfr.201900257.

[92] Xie, Chencheng, and Dina Halegoua-DeMarzio. 2019. "Role of Probiotics in Non-alcoholic Fatty Liver Disease: Does Gut Microbiota Matter?" *Nutrients* 11(11):E2837. doi: 10.3390/nu 11112837.

[93]   Tajiri, Kazuto, and Yukihiro Shimizu. 2018. "Branched-chain amino acids in liver diseases." *Translational Gastroenterology and Hepatology* 3:47. doi: 10.21037/tgh.2018.07.06.

[94]   Ahn, Sang Bong, Dae Won Jun, Bo-Kyeong Kang, Jong Hyun Lim, Sanghyun Lim, and Myung-Jun Chung. 2019. "Randomized, Double-blind, Placebo-controlled Study of a Multispecies Probiotic Mixture in Nonalcoholic Fatty Liver Disease." *Scientific Reports* 9(1):5688. doi: 10.1038/s41598-019-42059-3.

[95]   Duseja, Ajay, Subrat K Acharya, Manu Mehta, Shruti Chhabra, Shalimar, Satyavati Rana, Ashim Das, Siddhartha Dattagupta, Radha K Dhiman, and Yogesh K Chawla. 2019. "High potency multistrain probiotic improves liver histology in non-alcoholic fatty liver disease (NAFLD): a randomised, double-blind, proof of concept study." *BMJ Open Gastroenterology* 6(1):e000315. doi: 10.1136/bmjgast-2019-000315.

[96]   Pereira, Keith, Jason Salsamendi, and Javier Casillas. "The Global Nonalcoholic Fatty Liver Disease Epidemic: What a Radiologist Needs to Know." *Journal of Clinical Imaging Science* 5:32. doi: 10. 4103/2156-7514.157860.

[97]   Luther, Jay, John J. Garber, Hamed Khalili, Maneesh Dave, Shyam Sundhar Bale, Rohit Jindal, Daniel L. Motola, et al. 2015. "Hepatic Injury in Nonalcoholic Steatohepatitis Contributes to Altered Intestinal Permeability." *Cellular and Molecular Gastroenterology and Hepatology* 1(2):222-32. doi: 10.1016/j.jcmgh.2015.01.001.

[98]   Benedict, Mark, and Xuchen Zhang. 2017. "Non-alcoholic fatty liver disease: An expanded review," *World Journal of Hepatology* 9(16):715-32. doi: 10.4254/wjh.v9.i16.715.

[99]   Dukowicz, Andrew C., Brian E. Lacy, and Gary M. Levine. 2007. "Small intestinal bacterial overgrowth: A comprehensive review." *Gastroenterology and Hepatology* 3(2):112–22. https://www.ncbi. nlm.nih.gov/pmc/articles/PMC3099351/.

[100]   Bibbò, Stefano, Gianluca Ianiro, Maria Pina Dore, Claudia Simonelli, Estelle E. Newton, and Giovanni Cammarota. 2018. "Gut Microbiota as a Driver of Inflammation in Nonalcoholic Fatty Liver Disease."

Mediators of Inflammation 2018:9321643. doi: 10.1155/2018/9321643.

[101] Grabherr, Felix, Christoph Grander, Maria Effenberger, Timon Erik Adolph, and Herbert Tilg. 2019. "Gut Dysfunction and Non-alcoholic Fatty Liver Disease." *Frontiers in Endocrinology* 10:611. doi: 10.3389/fendo.2019.00611.

[102] Fitriakusumah, Yoga, C. Rinaldi A. Lesmana, Winda Permata Bastian, Chyntia O. M. Jasirwan, Irsan Hasan, Marcellus Simadibrata, Juferdy Kurniawan, Andri Sanityoso Sulaiman, and Rino A. Gani. 2019. "The role of Small Intestinal Bacterial Overgrowth (SIBO) in Non-alcoholic Fatty Liver Disease (NAFLD) patients evaluated using Controlled Attenuation Parameter (CAP) Transient Elastography (TE): a tertiary referral center experience." *BMC Gastroenterology* 19(1):43. doi: 10.1186/s12876-019-0960-x.

[103] Boursier, Jérôme, Olaf Mueller, Matthieu Barret, Mariana Machado, Lionel Fizanne, Felix Araujo-Perez, Cynthia D. Guy, et al. 2016. "The severity of nonalcoholic fatty liver disease is associated with gut dysbiosis and shift in the metabolic function of the gut microbiota." *Hepatology* 63(3):764-75. doi: 10.1002/hep.28356.

[104] Del Chierico, Federica, Valerio Nobili, Pamela Vernocchi, Alessandra Russo, Cristiano De Stefanis, Daniela Gnani, Cesare Furlanello, et al. 2017. "Gut microbiota profiling of pediatric nonalcoholic fatty liver disease and obese patients unveiled by an integrated meta-omics-based approach." *Hepatology* 65(2):451-64. doi: 10.1002/hep.28572.

[105] Aragonès, Gemma, Marina Colom-Pellicer, Carmen Aguilar, Esther Guiu-Jurado, Salomé Martínez, Fàtima Sabench, José Antonio Porras, et al. 2020. "Circulating microbiota-derived metabolites: a "liquid biopsy?" *International Journal of Obesity* 44(4):875-85. doi: 10.1038/s41366-019-0430-0.

[106] Kolodziejczyk, Aleksandra A, Danping Zheng, Oren Shibolet, and Eran Elinav. 2019. "The role of the microbiome in NAFLD and NASH." *EMBO Molecular Medicine* 11(2):e9302. doi: 10.15252/emmm.201809302.

[107] Xie, Guoxiang, Xiaoning Wang, Ping Liu, Runmin Wei, Wenlian Chen, Cynthia Rajani, Brenda Y. Hernandez, et al. 2016. "Distinctly altered gut microbiota in the progression of liver disease." *Oncotarget* 7(15):19355-66. doi: 10.18632/oncotarget.8466.

[108] Qian, Minyi, Haiyang Hu, Ying Yao, Danyang Zhao, Shilei Wang, Chuyue Pan, Xubin Duan, et al. 2020. "Coordinated changes of gut microbiome and lipidome differentiates nonalcoholic steatohepatitis (NASH) from isolated steatosis." *Liver International* 40(3):622-37. doi: 10.1111/liv.14316.

[109] Duparc, Thibaut, Hubert Plovier, Vannina G Marrachelli, Matthias Van Hul, Ahmed Essaghir, Marcus Ståhlman, Sébastien Matamoros, et al. 2017. "Hepatocyte MyD88 affects bile acids, gut microbiota and metabolome contributing to regulate glucose and lipid metabolism." *Gut* 66(4):620-32. doi: 10.1136/gutjnl-2015-310904.

[110] Akiyama, Kentaro, Eiji Warabi, Kosuke Okada, Toru Yanagawa, Tetsuro Ishii, Katsumi Kose, Katsutoshi Tokushige, et al. 2018. "Deletion of both p62 and Nrf2 spontaneously results in the development of nonalcoholic steatohepatitis." *Experimental Animals* 67(2):201-18. doi: 10.1538/expanim.17-0112.

[111] Bray, Freddie, Jacques Ferlay, Isabelle Soerjomataram, Rebecca L. Siegel, Lindsey A. Torre, and Ahmedin Jemal. 2018. "Global cancer statistics 2018: GLOBOCAN estimates of incidence and mortality worldwide for 36 cancers in 185 countries." *CA: A Cancer Journal for Clinicians* 68(6):394–424. doi: 10.3322/caac.21492.

[112] Forner, Alejandro, María Reig, and Jordi Bruix. 2018. "Hepatocellular carcinoma." *The Lancet* 391(10127):1301–14. doi: 10.1016/S0140-6736(18)30010-2.

[113] Ni, Jiajia, Rong Huang, Huifang Zhou, Xiaoping Xu, Yang Li, Peihua Cao, Kebo Zhong, et al. 2019. "Analysis of the relationship between the degree of dysbiosis in gut microbiota and prognosis at different stages of primary hepatocellular carcinoma." *Frontiers in Microbiology* 10:1458. doi: 10.3389/fmicb.2019.01458.

[114] Ponziani, Francesca Romana, Sherrie Bhoori, Chiara Castelli, Lorenza Putignani, Licia Rivoltini, Federica Del Chierico, Maurizio

Sanguinetti, et al. 2019. "Hepatocellular Carcinoma Is Associated With Gut Microbiota Profile and Inflammation in Nonalcoholic Fatty Liver Disease." *Hepatology* 69(1):107–20. doi: 10.1002/hep. 30036.

[115] Liu, Qisha, Fan Li, Yaoyao Zhuang, Jian Xu, Jianwei Wang, Xuhua Mao, Yewei Zhang, and Xingyin Liu. 2019. "Alteration in gut microbiota associated with hepatitis B and non-hepatitis virus related hepatocellular carcinoma." *Gut Pathogens* 11:1. doi: 10.1186/s13099-018-0281-6.

[116] Zheng, Ruipeng, Guoqiang Wang, Zhiqiang Pang, Nan Ran, Yinuo Gu, Xuewa Guan, Yuze Yuan, et al. 2020. "Liver cirrhosis contributes to the disorder of gut microbiota in patients with hepatocellular carcinoma." *Cancer Medicine.* doi: 10.1002/cam4. 3045.

[117] Méndez-sánchez, Nahum, Alejandro Valencia-Rodríguez, Alfonso Vera-Barajas, Ludovico Abenavoli, Emidio Scarpellini, Guadalupe Ponciano-Rodríguez, and David Q.-H. Wang. 2020. "The Mechanism of Dysbiosis in Alcoholic Liver Disease Leading to Liver Cancer." *Hepatoma Research* 6(5). https://hrjournal.net/ article/view/3350.

[118] Raza, Sana, Sangam Rajak, Baby Anjum, and Rohit A. Sinha. 2019. "Molecular links between non-alcoholic fatty liver disease and hepatocellular carcinoma." *Hepatoma Research* 5:42. doi: 10.20517/2394-5079.2019.014.

[119] Takakura, Kazuki, Tsunekazu Oikawa, Masanori Nakano, Chisato Saeki, Yuichi Torisu, Mikio Kajihara, and Masayuki Saruta. 2019. "Recent insights into the multiple pathways driving non-alcoholic steatohepatitis-derived hepatocellular carcinoma." *Frontiers in Oncology* 9:762. doi: 10.3389/fonc.2019.00762.

[120] Mohamadkhani, Ashraf. 2018. "On the potential role of intestinal microbial community in hepatocarcinogenesis in chronic hepatitis B." *Cancer Medicine* 7(7):3095–100. doi: 10.1002/cam4.1550.

[121] Schwabe, Robert F., and Tim F. Greten. 2020. "Gut microbiome in HCC – Mechanisms, diagnosis and therapy." *Journal of Hepatology* 72(2):230–8. doi: 10.1016/j.jhep.2019.08.016.

[122] Wu, Jian. 2016. "Utilization of animal models to investigate nonalcoholic steatohepatitis-associated hepatocellular carcinoma." *Oncotarget* 7(27):42762–76. doi: 10.18632/oncotarget.8641.

[123] Liew, Winnie-Pui-Pui, and Sabran Mohd-Redzwan. 2018. "Mycotoxin: Its impact on gut health and microbiota." *Frontiers in Cellular and Infection Microbiology* 8:60. doi: 10.3389/fcimb.2018. 00060.

[124] Yang, Wan-Shui, Xu-Fen Zeng, Zhi-Ning Liu, Qi-Hong Zhao, Yu-Ting Tan, Jing Gao, Hong-Lan Li, and Yong-Bing Xiang. 2020. "Diet and liver cancer risk: A narrative review of epidemiologic evidence." *British Journal of Nutrition* 1-11. doi: 10.1017/S0007114 520001208.

[125] Poutahidis, Theofilos, Bernard J. Varian, Tatiana Levkovich, Jessica R. Lakritz, Sheyla Mirabal, Caitlin Kwok, Yassin M. Ibrahim, et al. 2015. "Dietary microbes modulate transgenerational cancer risk." *Cancer Research* 75(7):1197–204. doi: 10.1158/0008-5472.CAN-14-2732.

[126] Rong, Yihui, Zheng Dong, Zhixian Hong, Yun Jin, Wei Zhang, Bailong Zhang, Wei Mao, et al. 2017. "Reactivity toward *Bifidobacterium longum* and *Enterococcus hirae* demonstrate robust CD8[+] T cell response and better prognosis in HBV-related hepatocellular carcinoma." *Experimental Cell Research* 358(2):352–9. doi: 10.1016/j.yexcr.2017.07.009.

[127] Pedersen, Kim B, Casey F Pulliam, Aarshvi Patel, Fabio Del Piero, Tatiane T N Watanabe, Umesh D Wankhade, Kartik Shankar, Chindo Hicks, and Martin J Ronis. 2019. "Liver tumorigenesis is promoted by a high saturated fat diet specifically in male mice and is associated with hepatic expression of the proto-oncogene Agap2 and enrichment of the intestinal microbiome with Coprococcus." *Carcinogenesis* 40(2):349–59. doi: 10.1093/carcin/bgy141.

[128] Huang, Rong, Ting Li, Jiajia Ni, Xiaochun Bai, Yi Gao, Yang Li, Peng Zhang, and Yan Gong. 2018. "Different Sex-Based Responses of Gut Microbiota during the Development of Hepatocellular Carcinoma in Liver-Specific *Tsc1*-Knockout Mice." *Frontiers in Microbiology* 9:1008. doi: 10.3389/fmicb.2018.01008.

[129] Xie, Guoxiang, Xiaoning Wang, Aihua Zhao, Jingyu Yan, Wenlian Chen, Runqiu Jiang, Junfang Ji, et al. 2017. "Sex-dependent effects on gut microbiota regulate hepatic carcinogenic outcomes." *Scientific Reports* 7:45232. doi: 10.1038/srep45232.

[130] Grąt, M., K. M. Wronka, M. Krasnodębski, Ł. Masior, Z. Lewandowski, I. Kosinska, K. Grąt, et al. 2016. "Profile of Gut Microbiota Associated With the Presence of Hepatocellular Cancer in Patients With Liver Cirrhosis." *Transplantation Proceedings* 48(5):1687–91. doi: 10.1016/j.transproceed.2016.01.077.

[131] Piñero, Federico, Martín Vazquez, Patricia Baré, Cristian Rohr, Manuel Mendizabal, Mariela Sciara, Cristina Alonso, Fabián Fay, and Marcelo Silva. 2019. "A different gut microbiome linked to inflammation found in cirrhotic patients with and without hepatocellular carcinoma." *Annals of Hepatology* 18(3):480–7. doi: 10.1016/j.aohep.2018.10.003.

[132] Zhang, Lei, Yong-Na Wu, Tuo Chen, Cheng-Hui Ren, Xun Li, and Guang-Xiu Liu. 2019. "Relationship between intestinal microbial dysbiosis and primary liver cancer." *Hepatobiliary and Pancreatic Diseases International* 18(2):149–57. doi: 10.1016/j.hbpd.2019.01. 002.

[133] Zhen, Hongde, Xiang Qian, Xiaoxuan Fu, Zhuo Chen, Aiqin Zhang, and Lei Shi. 2019. "Regulation of Shaoyao Ruangan Mixture on Intestinal Flora in Mice with Primary Liver Cancer." *Integrative Cancer Therapies* 18:1534735419843178. doi: 10.1177/153473541 9843178.

[134] Han, Xiao, Tianren Huang, and Junqing Han. 2019. "Cytokines derived from innate lymphoid cells assist *Helicobacter hepaticus* to aggravate hepatocellular tumorigenesis in viral transgenic mice." *Gut Pathogens* 11:23. doi: 10.1186/s13099-019-0302-0.

[135] Goossens, Jean-François, and Christian Bailly. 2019. "Ursodeoxycholic acid and cancer: From chemoprevention to chemotherapy." *Pharmacology and Therapeutics* 203:107396. doi: 10.1016/j.pharmthera.2019.107396.

[136] Sheng, Lili, Prasant Kumar Jena, Ying Hu, Hui-Xin Liu, Nidhi Nagar, Karen M Kalanetra, Samuel William French, Samuel Wheeler French, David A Mills, and Yu-Jui Yvonne Wan. 2017. "Hepatic inflammation caused by dysregulated bile acid synthesis is reversible by butyrate supplementation." *The Journal of Pathology* 243(4):431–41. doi: 10.1002/path.4983.

[137] Yoshimoto, Shin, Tze Mun Loo, Koji Atarashi, Hiroaki Kanda, Seidai Sato, Seiichi Oyadomari, Yoichiro Iwakura, et al. 2013. "Obesity-induced gut microbial metabolite promotes liver cancer through senescence secretome." *Nature* 499(7456):97-101. doi: 10.1038/nature12347.

[138] Loo, Tze Mun, Fumitaka Kamachi, Yoshihiro Watanabe, Shin Yoshimoto, Hiroaki Kanda, Yuriko Arai, Yaeko Nakajima-Takagi, et al. 2017. "Gut microbiota promotes obesity-associated liver cancer through PGE$_2$-mediated suppression of antitumor immunity." *Cancer Discovery* 7(5):522–38. doi: 10.1158/2159-8290.CD-16-0932.

[139] Wu, Rihui, Xueting Mei, Yibiao Ye, Ting Xue, Jiasheng Wang, Wenjia Sun, Caixia Lin, Ruoxue Xue, Jiabao Zhang, and Donghui Xu. 2019. "Zn(II)-curcumin solid dispersion impairs hepatocellular carcinoma growth and enhances chemotherapy by modulating gut microbiota-mediated zinc homeostasis." *Pharmacological Research* 150:104454. doi: 10.1016/j.phrs.2019.104454.

[140] Zhang, Xiao, Lei Wu, Yanquan Xu, Hua Yu, Yu Chen, Huakan Zhao, Juan Lei, et al. 2020. "Microbiota-derived SSL6 enhances the sensitivity of hepatocellular carcinoma to sorafenib by down-regulating glycolysis." *Cancer Letters* 481:32-44. doi: 10.1016/j.canlet.2020.03.027.

[141] Li, Jun, Cecilia Ying Ju Sung, Nikki Lee, Yueqiong Ni, Jussi Pihlajamäki, Gianni Panagiotou, and Hani El-Nezami. 2016. "Probiotics modulated gut microbiota suppresses hepatocellular carcinoma growth in mice." *Proceedings of the National Academy of Sciences of the United States of America* 113(9):E1306–15. doi: 10.1073/pnas.1518189113.

[142] Pant, Kishor, Anoop Saraya, and Senthil K. Venugopal. 2017. "Oxidative stress plays a key role in butyrate-mediated autophagy via Akt/mTOR pathway in hepatoma cells." *Chemico-Biological Interactions* 273:99–106. doi: 10.1016/j.cbi.2017.06.001.

[143] Pant, Kishor, Ajay K. Yadav, Parul Gupta, Rakibul Islam, Anoop Saraya, and Senthil K. Venugopal. 2017. "Butyrate induces ROS-mediated apoptosis by modulating miR-22/SIRT-1 pathway in hepatic cancer cells." *Redox Biology* 12:340-9. doi: 10.1016/j.redox.2017.03.006.

[144] Singh, Vishal, Beng San Yeoh, Benoit Chassaing, Xia Xiao, Piu Saha, Rodrigo Aguilera Olvera, John D. Lapek Jr, et al. 2018. "Dysregulated Microbial Fermentation of Soluble Fiber Induces Cholestatic Liver Cancer." *Cell* 175(3):679-94.e22. doi: 10.1016/j.cell.2018.09.004.

[145] Ejima, Chieko, Haruna Kuroda, and Sonoko Ishizaki. 2016. "A novel diet-induced murine model of steatohepatitis with fibrosis for screening and evaluation of drug candidates for nonalcoholic steatohepatitis." *Physiological Reports* 4(21):e13016. doi: 10.14814/phy2.13016.

[146] Yamada, Shoji, Masaki Kimura, Yoshimasa Saito, and Hidetsugu Saito. 2018. "Nrf2-mediated anti-oxidant effects contribute to suppression of non-alcoholic steatohepatitis-associated hepatocellular carcinoma in murine model." *Journal of Clinical Biochemistry and Nutrition* 63(2):123–8. doi: 10.3164/jcbn.17-125.

[147] Yamada, Shoji, Yoko Takashina, Mitsuhiro Watanabe, Ryogo Nagamine, Yoshimasa Saito, Nobuhiko Kamada, and Hidetsugu Saito. 2018. "Bile acid metabolism regulated by the gut microbiota promotes non-alcoholic steatohepatitis-associated hepatocellular carcinoma in mice." *Oncotarget* 9(11):9925-39. doi: 10.18632/oncotarget.24066.

[148] Shalapour, Shabnam, Xue-Jia Lin, Ingmar N. Bastian, John Brain, Alastair D. Burt, Alexander A. Aksenov, Alison F. Vrbanac, et al. 2017. "Inflammation-induced IgA+ cells dismantle anti-liver cancer immunity." *Nature* 551(7680):340-5. doi: 10.1038/nature24302.

[149] Singh, Vishal, Beng San Yeoh, Ahmed A. Abokor, Rachel M. Golonka, Yuan Tian, Andrew D. Patterson, Bina Joe, Mathias Heikenwalder, and Matam Vijay-Kumar. 2020. "Vancomycin prevents fermentable fiber-induced liver cancer in mice with dysbiotic gut microbiota." *Gut Microbes* 1-15. doi: 10.1080/19490976.2020.1743492.

[150] Iida, Noriho, Eishiro Mizukoshi, Tatsuya Yamashita, Takeshi Terashima, Kuniaki Arai, Jun Seishima, and Shuichi Kaneko. 2019. "Overuse of antianaerobic drug is associated with poor postchemotherapy prognosis of patients with hepatocellular carcinoma." *International Journal of Cancer* 145(10):2701–11. doi: 10.1002/ijc.32339.

[151] Yang, Baiyu, Katrina Wilcox Hagberg, Jie Chen, Vikrant V Sahasrabuddhe, Barry I Graubard, Susan Jick, and Katherine A McGlynn. 2016. "Associations of antibiotic use with risk of primary liver cancer in the Clinical Practice Research Datalink." *British Journal of Cancer* 15(1):85-9. doi: 10.1038/bjc.2016.148.

[152] Cassano, Marco, Sandra Offner, Evarist Planet, Alessandra Piersigilli, Suk Min Jang, Hugues Henry, Markus B. Geuking, et al. 2017. "Polyphenic trait promotes liver cancer in a model of epigenetic instability in mice." *Hepatology* 66(1):235–51. doi: 10.1002/hep.29182.

[153] Ma, Chi, Miaojun Han, Bernd Heinrich, Qiong Fu, Qianfei Zhang, Milan Sandhu, David Agdashian, et al. 2018. "Gut microbiome–mediated bile acid metabolism regulates liver cancer via NKT cells." *Science* 360(6391):eaan5931. doi: 10.1126/science.aan5931.

[154] Chi, Liang, Jingchuan Xue, Pengcheng Tu, Yunjia Lai, Hongyu Ru, and Kun Lu. 2019. "Gut microbiome disruption altered the biotransformation and liver toxicity of arsenic in mice." *Archives of Toxicology* 93(1):25–35. doi: 10.1007/s00204-018-2332-7.

[155] Aguilar-Olivos, Nancy E., Paloma Almeda-Valdes, Carlos A. Aguilar-Salinas, Misael Uribe, and Nahum Méndez-Sánchez. 2016. "The role of bariatric surgery in the management of nonalcoholic fatty

liver disease and metabolic syndrome." *Metabolism: Clinical and Experimental* 65(8):1196-207. doi: 10.1016/j.metabol.2015. 09.004.

[156] Tan, Chun Hai, Nawaf Al-Kalifah, Kong-Han Ser, Yi-Chih Lee, Jung-Chien Chen, and Wei-Jei Lee. 2018. "Long-term effect of bariatric surgery on resolution of nonalcoholic steatohepatitis (NASH): An external validation and application of a clinical NASH score." *Surgery for Obesity and Related Diseases* 14(10):1600-6. doi: 10.1016/j.soard.2018.05.024.

[157] Nobili, Valerio, Antonella Mosca, Tommaso Alterio, Sabrina Cardile, and Lorenza Putignani. 2019. "Fighting Fatty Liver Diseases with Nutritional Interventions, Probiotics, Symbiotics, and Fecal Microbiota Transplantation (FMT)." *Advances in Experimental Medicine and Biology* 1125:85-100. doi: 10.1007/ 5584_2018_318.

[158] Vaughn, Byron P., Kevin M. Rank, and Alexander Khoruts. 2019. "Fecal Microbiota Transplantation: Current Status in Treatment of GI and Liver Disease." *Clinical Gastroenterology and Hepatology* 17(2):353-61. doi: 10.1016/j.cgh.2018.07.026.

[159] Xue, Lan-Feng, Wen-Hui Luo, Li-Hao Wu, Xing-Xiang He, Harry Hua-Xiang Xia, and Yu Chen. 2019. "Fecal Microbiota Transplantation for the Treatment of Nonalcoholic Fatty Liver Disease." *Exploratory Research and Hypothesis in Medicine* 4(1):12-18. doi: 10.14218/ERHM.2018.00025.

In: Dysbiosis: A Study of Underlying Causes   ISBN: 978-1-53618-332-0
Editor: Richard I. Cowell   © 2020 Nova Science Publishers, Inc.

Chapter 3

ORAL MICROBIAL DYSBIOSIS

Nursen Topcuoglu and Guven Kulekci
Istanbul University Faculty of Dentistry,
Department of Oral Microbiology, Istanbul, Turkey

ABSTRACT

The oral cavity, which has more than 700 bacterial species on teeth and mucosal surfaces, houses one of the most diverse microbial community in the human body. Distinct microenvironments containing heterogeneous microbes in the mouth form an important link for mouth and general health.

The microorganisms that make up the microbiome form highly regulated and well organized communities attached to surfaces as biofilms, which contribute to ecologic equilibrium. Substantial changes in a local environment can alter the competitiveness of biofilm bacteria may disturb this equilibrium, leading to enrichment of organisms most suited to the new environment, which can be defined as dysbiosis. The vast majority of oral diseases such as caries and gingivitis are caused by the ecological shift in oral biofilms.

Various genetic, epigenetic or patient-modifiable factors contributes to dysbiosis: such as immunological diseases, hormonal disorders, puberty, diabetes, stress, eating habits, smoking, antibiotic/antimicrobial agent use, poor oral hygiene and gingivitis cause changes in symbiotic bacterial community.

It is important to promote a balanced microbiome to maintain or improve oral health effectively. In order to fully understand the progression of oral diseases, it is essential to evaluate the microbiome together with genetic, immunological and environmental factors.

Keywords: oral microbiota, dysbiosis, oral biofilms

INTRODUCTION

The oral microbiome, one of the important parts of the human microbiome, consist of a highly diverse range of microorganisms which are mostly indigenous. The diversity depends on the wide variety of surfaces in the oral cavity for bacterial attachment and colonization. These surfaces include the mineralized hard tissues of the teeth along with the soft tissues of the oral mucosa, cheeks, gums and tongue. They continuously bathed in saliva or gingival crevicular fluid (GCF), and so, provides a nutritious and warm habitat, with a prevailing pH and a range of gaseous atmospheric conditions that are suitable for the growth of a wide range of microbial genera [1]. The microbiota exists in the mouth as multispecies biofilms, the composition and metabolic activity of which is determined by host and environmental factors [2].

There is a dynamic relationship between the host, the environment and the resident microbiota [3]. Like elsewhere in the body, the oral microbiome has a symbiotic relationship with the host. A breakdown of the symbiotic relationship between the oral microbiome and the host (dysbiosis), results in oral diseases like caries and periodontitis [4, 5].

In this chapter, the dynamic relationship between the host, the environment and the resident microbiota and the role of microbiota in oral diseases has been reviewed extensively.

Oral Ecosystems

The oral cavity provides a variety of ecologically distinct surfaces for bacterial attachment and colonization. These include the mineralized hard tissues of the teeth along with the soft tissues of the oral mucosa, cheeks, gums and tongue. Tooth surface over gingival called as *supragingival* region; the area between the root and the sulcular epithelium under the gums is called as *subgingival* region. All that distinct habitats support the growth of a characteristic microbial community because of their particular biological and physical features [3]. The hard and non-shedding feature of the teeth, provides the opportunity for substantial biofilm formation. Saliva and GCF contribute to the acquired pellicle formation on teeth, and are also the primary sources of nutrients for the microorganisms. In contrast, they help to regulate bacterial and fungal colonization by flushing and removing the weakly-attached microorganisms as well as delivering the innate and adaptive immune system components. Tissue-specific tropisms are often defined by specificity and avidity of adherence, which is a feature of many successful oral colonizers, providing resistance to the mechanical shearing forces of fluid flow and mastication [6, 7].

The factors that influence the growth of microorganisms in the mouth include the redox potential, pH and temperature of the sites, the activity of the host defenses and the nutrition and antimicrobial agent usage by the host.

Oral Microbiota

Developments in Historical Order

Oral microbiota refers to the microorganisms found in the human oral cavity. It was first identified by the Antony van Leeuwenhoek, the pioneer of modern microbiology, using a microscope constructed by him. The microbes sketched in his notebook during the mid-1670s are now known as some of the most abundant bacteria resided within oral cavity, including cocci, spirochetes, and fusiform bacteria. W. D. Miller was arguably one of the most important individuals who greatly advanced oral microbiology. He studied at Robert Koch's laboratory for the purpose of identifying the "germs" that were responsible for tooth decay. In his book titled "Microorganisms of the Human Mouth" written in 1890, he proposed the "chemoparasitic" theory which was provided the key elements for our modern concept of the etiology of dental caries [8]. In 1924, J. K. Clarke first isolated a bacterial species from human dental caries site, it was named *Streptococcus mutans*, and was shown to be capable of fermenting several sugars and producing a pH of 4.2 in glucose broth [9]. Fitzgerald and Keyes [10] demonstrated this bacterium was the etiological agent of dental caries in 1960. The improved anaerobic cultivation techniques allowed successful isolation and characterization of microbial species from subgingival dental plaque, which naturally prompted oral microbiologists to connect specific bacterial species to periodontal diseases.

Recombinant DNA methodology developed in the late 1970s and genetic mapping, has been used to genetically manipulate and characterize the microorganisms. The advent of culture-independent methods has greatly improved the detection of microorganisms, many of which cannot yet be grown in culture. The most common culture-independent technique to analyze the microbiome is based on 16S ribosomal RNA (16S rRNA) gene community profiling [11]. The 16S rRNA gene is present in all prokaryotes and contains variable regions that are unique between microorganisms and that can be used as a means of identification.

The genetic studies, together with the additional bio-informatic information obtained from whole genomic sequences of many bacteria allow microbiologists to dissect the development and function of microbial communities at the molecular level. The traditional method of 16S rRNA gene sequencing was costly, laborious and time-consuming. The advent of Next Generation Sequencing (NGS) methods such as 454 pyrosequencing and Illumina MiSeq have enabled a massively increased sample throughput, with up to 27 million sequences being generated in a single run (compared with a few hundred with the traditional method) [12]. This method is a useful tool that allows for high-volume studies of the genetic material in samples and has greatly increased our knowledge and understanding of the oral microbiome.

Human Microbiome Project and Human Oral Microbiome Database

The oral cavity supports a wide range of microorganisms, predominated by bacteria. The distinct physicochemical features of the different microenvironments of the host and tissue-specific tropisms of the microorganisms leads successful oral colonization. The spatial and structural organization of natural microbial communities is essential for physical and metabolic interspecies interactions that can be both antagonistic or cooperative [6, 7].

To understand the microbial communities in heath and disease, Human Microbiome Project (HMP) was established by National Institute of Health (NIH) Common Fund, in 2008. Oral cavity was one of the five major body area nearby nasal cavity, vagina, intestinal tract and skin. The bacteria of oral that have been sequenced accounted for 26% of all the body sites [13,14]. The oral cavity taxon, which was one of less dominant taxa, was suggested to be highly personalized. As saliva, buccal mucosa (cheek), keratinized gingiva, palate, tonsils, throat and tongue soft tissues, and supragingival and subgingival dental plaque were investigated in the oral cavity, most habitats were dominated by *Streptococcus,* but these were followed in abundance by *Haemophilus* in the buccal mucosa, *Actinomyces* in the supragingival plaque, and *Prevotella* in the immediately adjacent (but low oxygen) subgingival plaque. The study also revealed relationship among microbial community membership and function [13].

Sequencing of the bacterial 16S rRNA gene has yielded invaluable information on the bacterial components of the oral microbiome and has allowed for the creation of the Human Oral Microbiome Database (HOMD) [15]. By addition of 84 microbial species on the microbiota of the aero digestive tract outside of the mouth, the expended version (*e*HOMD) includes a total of 789 microbial species, which 57% are officially named, 13% unnamed but cultivated and 30% are known only as uncultivated phylotypes [16]. It presents a provisional naming scheme for the currently unnamed taxa, based on the 16S rRNA sequence phylogeny, so that strain, clone and probe data from any laboratory can be directly linked to a stably named reference scheme [17]. A total of 1,570 genomes, representing 475 taxa (62% of all taxa, 85% of cultivated taxa) are currently available on *e*HOMD which links sequence data with phenotypic, phylogenetic, clinical and bibliographic information [18].

Totally 16 bacterial phyla, displaying by *e*HOMD are (in alphabetical order): Absconditabacteria (SR1), Actinobacteria, Bacteroidetes, Chlamydiae, Chlorobi, Chloroflexi, Cyanobacteria, Euryarchaeota, Firmicutes, Fusobacteria, Gracilibacteria (GN02), Proteobacteria, Saccharibacteria (TM7), Spirochaetes, Synergistetes and WPS-2; which all represents 231 genera and 789 species [18].

Acquisition of Oral Microbiota

The oral cavity generally lacks of significant bacterial colonization at birth. In the first few minutes after birth the neonatal oral cavity has been shown to harbor microbiota resembling vaginal or skin microbiota, depending on mode of delivery, and continually contaminated by the bacteria from animate and inanimate objects [19]. The majority of the bacteria are transients. However, the bacteria obtained from exogenous saliva of their mothers and other individuals with close contact, mostly colonize the oral cavity in success.

The infants acquire an oral microbiota on the mucosa that after some weeks consists generally of *Streptococcus*, *Veillonella*, and *Neisseria* species [20]. The bacteria which produce immunoglobulin A (IgA) proteases such as *Streptococcus mitis* and *Streptococcus oralis,* are able to colonize the mucosal surfaces earlier [21]. The oral microbiota continues to develop, changing with age in composition and overall activity.

The early setting commensals provide an ecological advantage over later species and are selective in the placement of other bacteria as a result of their metabolism. Products such as lactic acid produced by *Streptococcus* species that metabolize oligosaccharides in breast milk by colonizing in the mouths of newborns easily by clinging to mucosal surfaces may cause bacterial species such as *Veillonella* to settle in the oral cavity [22]. Shorter breastfeeding habits and antibiotic treatment during the first 2 years of age are associated with a distinct bacterial composition at later age. In a longitudinal study, *Streptococcus* species decreased more with age in children with less breastfed and *Veillonella* species were higher in seven-year-old children who were breastfed up to 12 months compared to those who did not [23]. The introduction of new nutrient sources, is associated with increasing complexity for the first year of life [24].

Colonization by *Streptococcus sanguinis* is dependent on the presence of teeth at around six months. Oral streptococci and *Actinomyces* comprise a significant proportion of the organisms in dental plaque biofilm [21]. Colonization by *Streptococcus mutans*, the principal agent associated with dental caries, generally occurs by the deciduous molar eruption meanly in case of decrease in *S. sanguinis*. *S. sanguinis* group streptococci are antagonistic for *S. mutans*, due to their hydrogen peroxide production and their arginolytic effects, so they are mostly associated with health [25].

Gradual age-related changes may affect the composition of oral microbiota. Gram-negative anaerobic bacteria may increase during puberty, as a result of some hormones may act as nutritional sources for these bacteria. In adulthood, pregnancy or lifestyle events, such as smoking, frequency of carbohydrate consumption; in older, decline in general health and salivary flow rate or by tooth loss and usage of acrylic prosthesis, leads to changes in microbial composition. Differences among hosts, such as genetics and the (integrity of the) immune system also affect the microbiota composition.

Microbial succession is influenced not only by host factors, but also by microbial factors, as well. The metabolism of the aerobic and facultative anaerobic pioneer species lowers the redox potential in dental plaque biofilm and creates suitable conditions for colonization of strictly anaerobes. The metabolic end products of one organism may also be a nutrient source for another.

Oral Microbial Communities in Health

Once established, the oral microbiome is maintained by host- and microbe-derived factors. There is a symbiotic and a dynamic relationship with the host and the oral microbiome.

The host provides a nutritious habitat with various suitable conditions (oxygen concentrations, redox potential, pH, temperature, etc.) for colonization of a wide range of microorganisms, whereas, the oral microbiota delivers some key functions that provide important benefits to the host. The resident oral microbiota acts as a barrier to exogenous organisms (colonization resistance) and regulates the immune system [1]. Several beneficial effects like control the blood pressure or stimulation the gastric mucus production has also been shown [26, 27].

The microorganisms that make up the microbiome form highly regulated and well organized communities attached to surfaces as biofilms. Biofilms orderly and sequentially develop via a number of waves of microbial succession in which the diversity and richness of the microbiota increases over time [28]. The composition and metabolic activity is determined by host and environmental factors [2].

The early colonizers initially adhere to the tooth surface via weak electrostatic attractive forces, followed by a variety of specific molecular interactions between bacterial adhesins and receptors of salivary pellicle on the tooth surface. The synthesis of extracellular polysaccharides (EPS) from the bacteria contributes the plaque matrix and increase the probability of permanent attachment. Pioneer species modify the environment, enabling coadhesion between more fastidious species to attach and become established at a later time point. Microbial biofilms are both structurally and functionally organized and exist as highly interactive microbial communities which often participate in complementary metabolic interactions that ensure optimal substrate utilization [7]. Bacteria can also be antagonistic to another through competition for nutritional substrates and attachment sites and by the production of toxic metabolites and bacteriocins [21]. Bacteria can communicate with another through a chemical signaling system, *quarum sensing*: cell density-dependent regulation of gene expression. The system is mediated by a produced and secreted signaling molecule, called autoinducer (AI) Many oral organisms possess AI-2 signaling. Which can be expected to play a role in plaque development. Gram positive bacteria also signal through short peptides such as competence- stimulating peptide (CSP), involved in the process of DNA uptake and recombination [21].

A complex network of interdependencies exists among the members of the biofilm, and these contribute to maintaining community stability and resistance to change. Compared to the other microbial communities of the body, the oral microbiota in health is often considered most stable over time [29]. The continuous presence and the composition of saliva appear to have crucial roles in maintaining the stability of the oral microbiota [30]. The salivary microbiome was found to be less affected and more resilient toward the exposure to antibiotics than the fecal microbial community [31]. It is possible that, the oral microbial ecosystem possesses a higher intrinsic resilience toward stress, including recovery from exposure to antibiotics. This ecosystem has to surmount multiple daily perturbations such as oral hygiene measures, including exposure to topical antimicrobial agents and physical removal by tooth brushing, as well as alterations in temperature and oxygen [32].

Core Microbiome in Symbiosis

Core microbiome describes for least variable microbiota of a niche. It is shared with most of individuals and comprised of the predominant species in healthy conditions of oral cavity [11]. The dominance of Firmicutes, Bacteroidetes, Proteobacteria, Actinobacteria, Spirochaetes and Fusobacteria, constitutes 80% of total oral bacterial filum in a healthy oral cavity; where, Euryarchaeota, Chlamydia, Chloroflexi, SR1, Synergistetes, Tenericutes, Cyanobacteria, OD2, and TM7 also found in healthy subjects [14, 15, 33-36]. The most abundant genera include *Streptococcus, Prevotella, Neisseria, Haemophilus, Porphyromonas, Gemella, Rothia, Granulicatella, Fusobacterium, Actinomyces,* and *Veillonella* [37].

Oral bacterial microbiota is site-specific and different richness may be shown in different sites. Hard palate shows the lowest estimate of total richness, while the gingival plaque shows the highest estimate of total richness [38]. Firmicutes and Actinobacteria exhibit their abundance in dental plaques (Keisjer et al. 2008) while anaerobes (*Prevotella, Capnocytophaga, Flavobacterium*) of Bacteroidetes prefer niches of dorsal and lateral surfaces of the tongue and microaerophilic spaces [39].

Besides, fungi also constitute the significant proportion of an oral microbiota [40]. *Candida* spp. are one of the most common taxa of fungi that contribute to early in vitro biofilm formation [41]. The presence of Archaea is also have been reported from the oral cavity [42, 43].

In health, the majority of the bacteria have a symbiotic relationship with the host. The complex equilibrium between resident species in *symbiosis* in the oral cavity is responsible for the maintenance of a healthy state [12]. The composition of microbial communities (after the microbiome has matured in childhood) is remarkably [1, 44, 45], and more diverse stable in the healthy mouth [14, 46, 47].

In ecology, *resilience* is the capacity of an ecosystem to deal with perturbations without shifting to an alternative state in which core species and key functions are lost [30]. Resilience can be divided into "resistance" and "recovery": the resistance determines the magnitude of perturbation that an ecosystem can handle before its state changes, while the recovery is the rate at which it returns to its original state [30]. Healthy individuals can often adapt to the physiological changes, like age or hormonal changes in puberty and pregnancy, without detriment to their oral health [48].

Oral Microbiome in Disease: Dysbiosis

When disease drivers are strong or persistent enough, perturbations in oral ecology pass a certain threshold. As a consequence, the symbiotic relationship between the oral microbiome can breakdown and regime shifts (dysbiosis) take place in ecological systems which lead to oral diseases [30]. Modifiable factors driving oral dysbiosis include salivary gland dysfunction (changes in saliva flow and/or composition), diabetes, poor oral hygiene, gingival inflammation and lifestyle choices, including dietary habits and smoking [12, 49-52].

An "Ecological Plaque Hypothesis" has been proposed by Philip D. Marsh to explain the relationship between the resident oral microbiota and dental disease [53, 54]. In brief, the organisms associated with disease (pathobionts) may also be present in healthy sites, but at levels too low to be clinically relevant. The substantial change in local environmental conditions can alter the competitiveness of plaque bacteria leading to the enrichment of organisms most suited to the new environment. Disease is a result of a shift in the balance of the resident microbiota (dysbiosis) due to a response to a change in local environmental conditions. Pathobionts can grow to markedly higher proportions with elevated virulence potential than under healthy conditions. [44]. Although acute onset of both diseases can be triggered under particular host-compromising conditions, oral diseases such as periodontitis and dental caries are often chronic and slowly progressing [6].

A key principle of the 'Ecological Plaque Hypothesis' is that disease can be controlled not only by improving oral hygiene or targeting the putative pathogens directly, but also by interfering with the environmental pressures that select for the pathogenic micro-organisms, thereby driving dysbiosis [44].

Therefore, the knowledge of the mechanism of dysbiosis and the factors which cause dysbiosis is essential for the application of evidence-based therapy in oral diseases.

Dysbiosis as Origin of Dental Caries and Periodontitis

Dental Caries

Dental caries can be defined as the localized destruction of tooth tissues (demineralization) caused by acids produced from bacterial fermentation of dietary carbohydrates.

Early culture-based studies had shown that enamel caries was associated with increases in the numbers and proportions of mutans streptococci (MS), *Streptococcus mutans* and *Streptococcus sobrinus* [55], with lactobacilli being recovered from more advanced lesions [56]. However, while these bacteria can be seen on caries-free lesions, the absence of these bacteria on caries surfaces has also been reported [57-59].

Recent molecular analyses have strengthened this concept by showing that the microbiota associated with white spot lesions is more diverse than hitherto appreciated and that novel phylotypes and species including

Actinomyces spp, Scardovia wiggsiae and Slackia exigua as well as a broad range of non-MS, *Bifidobacterium* spp., *Veillonela* spp. and *Candida albicans* may also play a role [45, 58, 60-66].

Although there may be a lack of apparent specificity in the etiology of caries in terms of bacterial name, there is a definite specificity in terms of biochemical function [1]. The ability to rapidly convert dietary sugars to acid (acidogenicity) and, to be able to continue to grow and metabolize sugars under these acidic conditions (aciduricity) are the essential features of cariogenic microorganisms. In addition, mutans streptococci can also synthesize intracellular and extracellular polysaccharides (IPS and EPS, respectively) from sucrose. These capabilities not only provide a carbohydrate reserve that could be used to generate acid in the absence of dietary sugars, and makes a major contribution to the plaque matrix, as well.

Thus, excess fermentable carbohydrate consumption is one of the main factors for ecological shift. If sugar consumption is low and infrequent, despite the microorganisms are being able to produce acids that demineralize enamel, the episodic pH decrease can be readily neutralized by saliva, which restores and maintains remineralization, and ensures the microbial communities on teeth remain stable [67]. On the other hand, with frequent exposure to fermentable carbohydrates, microorganisms become embedded in an EPS-rich biofilm matrix while constantly producing acids that are physically protected from rapid buffering by saliva. Acidogenic species that are adapted to the acidic conditions will gain a selective advantage [54, 61, 68]. Over time, if the biofilm is not removed and frequent sugar consumption continues, the microbiota shifts toward an aciduric cariogenic microbiota, which are more adapted to growth and metabolism at a low pH [54]. A prolonged and repeated state of acidification disrupts the homeostatic mineral balance of the tooth surface in the direction of demineralization, resulting in dental caries.

Increased *Lactobacillus* levels, are correlated with a low pH (pH 4.5-5.0) [69]. Higher bacterial counts were observed in surface and superficial dentinal layers compared to deeper dentinal layers, whereas surface and shallow lesion layers were more acidic than deeper in dentin [70].

Caries microbiota can be less diverse than that of caries-free sites, suggesting suppression of the acid sensitive bacteria in the low pH of active lesions [62]. Lower microbial diversity was observed with increased dentin acidity [71]. The more acidic dentin had high proportions of *Lactobacillus* and *Atopobium* species whereas in less acidic lesions the microbiome was more diverse [71].

The abundance of Firmicutes shows the initiation of polysaccharide hydrolysis in the oral cavity. In an *in situ* study, a significant increase of the phylum Firmicutes (*Streptococcus* species) and a decrease of the phyla Proteobacteria (*Haemophilus* and *Aggregatibacter* species) and Bacteroidetes (*Porphyromonas* species) with a decrease in species richness, was observed after a 3-month dietary change of sucking 10 g sucrose per day in addition to the regular diet [72]. The results support the extended ecological plaque hypothesis and emphasize the synergy of multiple bacterial species in the development of caries.

When examining an ecosystem, it is important to know who the cells are, as well as what they are doing. Metatranscriptomics is simply looking at the transcriptome (the set of all *RNA* transcripts, for gene expression to make proteins) of many organisms at a time, such as all the bacteria in a plaque sample. Metabolomics is used to determine the metabolic activity of the microbiome and is another approach to examine what the bacteria are doing [73].

Metagenomic analysis indicated increased mixed-acid fermentations in caries compared with caries-free biofilms [74]. The acid can be destroyed by enzymes like urease and arginine deiminase, which produce ammonia and raise the biofilm pH, associated with health [75, 76].

Caries-free children and adults had shown to have higher arginine deiminase and lysine levels than those with caries [75, 77]. Ammonia production from arginine deiminase might be active in initial lesions, whereas urease activity might balance the acidic microbiota of more advanced disease [78].

According to current information, caries is a process involving the interactions of all microorganisms associated with disease and health with each other and with the host tissues and response systems [62]. We can use the analyze the composition and function of microbiome for risk assessment and treatment planning for dental caries.

Periodontal Diseases

"Periodontal diseases" is a connective term used to describe the inflammatory changes of tooth-supporting structures in which oral bacteria play an important role in the progress of disease.

In periodontal health, the gingiva surrounds the teeth at the level of the cementoenamel junction and forms gingival crevice from one to four millimeters deep [79]. Accumulation of dental biofilm as a result of poor oral hygiene, elicits a local inflammatory response at the area of gingival margin. This inflammatory lesion is termed gingivitis. Prolonged accumulation of dental biofilm causes longstanding gingival inflammation that may result in periodontitis: the gradual deepening of the gingival crevice (periodontal pocket) and in concomitant destruction of the periodontal ligament and the alveolar bone [79]. The disease progress depends on bacterial load, the composition of the microbial community, and host genetic factors. Thus, complex interactions between immune response mediators and the biofilm is important for disease progression from gingivitis to periodontitis. A dysbiotic microbial community subverts the host response so that most tissue damage is due to an inappropriate and uncontrolled level of inflammation [80, 81].

Local gingival inflammation causes an increased flow of the GCF and potentially bleeding, which whereby the site becomes deprived of oxygen, favoring the growth of anaerobic bacteria [44]. GCF proteins can act as a novel source of nutrients for proteolytic species that increase in number during periodontal diseases [53, 54]. As a result of the metabolism of asaccharolytic proteolytic species, the pH of the environment stays neutral or becomes slightly alkaline [82]. GCF and blood are important source for iron, which is essential for bacterial growth and triggers potential pathogenic mechanisms in oral bacteria associated with periodontal disease [83]. Therefore, inflammation tolerant, obligatory anaerobic, proteolytic, alkaliphilic species have a selective advantage and increase in number in the new environment [30, 53, 54, 68]. The dysbiotic microbiota induces the destruction of the periodontal tissue by a dysregulated host inflammatory immune response, which in turn provides new tissue breakdown-derived nutrients like degraded collagen and haem-containing compounds for the bacteria [4, 84].

In situ community-wide transcriptomic analysis of periodontitis-associated subgingival biofilms showed the elevated expression of proteolysis-related genes and genes for peptide transport and acquisition of iron and genes for the synthesis of lipopolysaccharides which increase pro-inflammatory response [85]. The addition of serum, hemoglobin or hemin to generated oral multispecies biofilms *in vitro* induces the selective outgrowth of organisms that can act as pathobionts, which additionally upregulate virulence genes including those encoding proteases, hemolysins and proteins involved in hemin transport [86]. Elevating the concentration of potassium enhanced production of pro-inflammatory cytokines and decreased production of human β-defensin 3 in gingival epithelial cells as a result of compositional and phenotypic changes in the microbial community was reported in an ex vivo dental plaque biofilm model [87]. Microbial dysbiosis was associated with the selective expansion of nitrate reductase-expressing Proteobacteria, which can use the elevated nitrate in the periodontal environment in a murine model [88]. The studies above, argues a reciprocal cause-and-effect relationship between dysbiosis and inflammation exists: inflammation fuels the selective growth of dysbiotic communities and dysbiosis exacerbates inflammation [6].

Recently, polymicrobial synergy and dysbiosis model ('PSD model') has been proposed by Hajishengallis and Lamont [83, 84] for periodontal disease pathogenesis. According to the PSD model, periodontitis-associated microbial communities show synergistic interactions for enhanced colonization, nutrient procurement, and persistence in an inflammatory environment. The dysbiosis of the periodontal microbiota signifies an imbalance in the relative abundance or influence of microbial species within the ecosystem (as compared to health), termed as *inflammophilic pathobionts,* leading to alterations in the host-microbial crosstalk sufficient to mediate destructive inflammation and bone loss [83, 84, 89].

The relationships among species were described by Socransky et al. [90]. Associations between all pairs of species were measured by various similarity coefficients and the resulting similarity matrices subjected to cluster analysis. These analyses supported the hypothesis that there were distinct complexes of microorganisms in subgingival biofilms. These analyses were valuable in that they confirmed the complexes described by the pair wise cluster analyses and suggested the nature of the relationships of the relationships of the complexes (communities) to each other [90]. Hence, these clusters included four complexes primarily consisting of early colonizers = blue complex: *Actinomyces* spp; yellow complex: *Streptococcus* spp; green complex: *Eikenella corrodens, Capnocytophaga* spp, *Campylobacter concisus* and *Aggregatibacter actinomycetemcomitans* sertype a; and purple complex: *Veillonella parvula* and *Actinomyces odontolyticus*. Two additional bacterial clusters intimately involved with pathological periodontal conditions are = orange complex: *Prevotella intermedi/nigrescens, Fusobacterium nucleatum, Fusobacterium periodonticum* and *Parvimonas micra*; red complex that are frequently elevated in plaque samples adjacent to periodontal lesions: *Porphyromonas gingivalis, Tannerella forsythia* and *Treponema denticola*. There are positive correlations within all the red complex bacteria and between red and orange complex bacteria [91]. Although strongly associated with localized aggressive periodontitis, *A. actinomycetemcomitans* serotype b did not cluster within any of the bacterial consortia.

Within the advent of culture-independent molecular studies, the list of candidate pathogens has extended to include the Gram-positive bacteria *Filifactor alocis* and *Peptoanaerobacter stomatis*; Gram-negative members of the Firmicutes phylum (*Dialister* spp., *Megasphaera* spp. and *Selenomonas* spp.); species in the genera *Prevotella, Desulfobulbus* and *Synergistes*; and many others [4, 14, 15, 89, 92-94]. Moreover, a plethora of virulence factors upregulated in the microbiome of periodontitis patients is primarily derived from species that are not traditionally considered as 'periopathogens' [85].

Microbiome diversity and, therefore, community complexity increases from health to periodontitis by a process of microbial succession, with the emergence of newly dominant species, but without replacement of pioneer health-associated species [89]. Health-associated species are those that occupy ~60 percent of the biomass in health and only ~10 percent of the biomass in periodontitis, while periodontitis-associated species increase their biomass ~4-log from health to periodontitis [4]. Total bacterial load in periodontal pockets are equal to 5 mm depth increases by at least 3-log from healthy pockets with <4 mm in depth [89]. The health associated genus *Actinomyces* was shown not to change its total biomass from health to disease. Therefore, its decreased relative proportion in periodontitis is the consequence of lack of growth as the whole community matures and increases in biomass [4]. In contrast, the total number of cells of the core genus *Veillonella*, did not change in terms of relative proportions from health to periodontitis, was shown to be higher in disease [89]. The core microbiome, which are probably capable of synergistic interactions with health- and disease associated species, increase their biomass from health to disease [4]. One of the most abundant core species, *Fusobacterium nucleatum*, can interact physically via coaggregation with a diverse range of oral species, has a positive influence on the biomass of the Gram-negative anaerobes, can easily adapt to aerated conditions and reduce the environment to anaerobic levels and generates CO_2, which is subsequently metabolized by *P. gingivalis* [95-97].

P. *gingivalis* is also present in health, albeit in lower numbers. The actual role of *P. gingivalis* involves its ability to initiate the conversion from a symbiotic community structure to a dysbiotic one capable of causing destructive inflammation [83]. In this regard, *P. gingivalis* expresses a variety of virulence factors (such as gingipains, atypical lipid A structures, and serine phosphatases), which manipulate the host response in ways that create a permissive environment for the growth of both *P. gingivalis* and bacteria co-habiting the same niche [83, 84]. This community-wide impact and the fact that, at early stages, *P. gingivalis* acts as a keystone pathogen that contributes to the dysbiotic process, leading to disease progression. Other members of red complex bacteria, *T. denticola* and *T. forsythia,* may contribute to the nososymbiocity of the microbial community once homeostasis is disrupted, thereby acting as pathobionts that accelerate disease progression. [6]. *P. gingivalis* is likely an important risk factor in periodontitis. By analogy to the crucial role of the literal keystone holding an entire arch together, *P. gingivalis* holds the members of the dysbiotic niche together [81, 83].

Oral Microbial Dysbiosis Related to Systematic Diseases

The pathogenesis of oral diseases is influenced by various host factors, including immune response or physiological factors, determined by the genetic profile of the host and may be modified by environmental and host behavioral factors. Certain systemic disorders share similar etiological factors especially with periodontal diseases, therefore, affected individuals may show manifestations of both diseases [98].

The vast majority of oral diseases such as caries and gingivitis are caused by the ecological shift in oral microbiota. The composition of the microbiota is governed by local factors, but systemic factors can also have a significant effect. The oral microbial profile is likely to shift, especially as a result of diseases related to immune system. For instance, Kostmann syndrome, also known as severe congenital neutropenia (SCN), includes a heterogeneous group of disorders characterized by chronic low absolute neutrophil counts in the peripheral blood. It results in early onset of bacterial infections, including early-onset periodontal breakdown [99, 100]. Higher bacterial load with lower bacterial diversity and differences in microbial profiles have been shown for the patients than the healthy controls [47, 101].

Individuals with Papillon-Lefèvre syndrome (PLS) develop severe gingival inflammation and pocket formation soon after eruption of teeth. The loss of periodontal attachment and alveolar bone progresses rapidly and leads to loss of the primary and permanent teeth at a young age [102]. The neutrophil deficiency of cathelicidin LL-37 peptide, compromise the host's ability to kill periodontal bacteria [103]. It has also been suggested that relentless recruitment and accumulation of hyperactive/reactive neutrophils in PLS causes the release of higher levels of proinflammatory cytokines, which together with reduced antimicrobial capacity of neutrophils, may lead to a locally destructive chronic inflammatory cycle [104].

Systemic diseases with increased inflammation are frequently linked to increased risk of periodontal disease [105]. Increased levels of periodontal disease, diabetes, systemic lupus erythematosus (SLE), and RA may reflect a common susceptibility, since each has inflammation as a common risk factor. Those individuals with a greater tendency toward inflammation have enhanced susceptibility and risk of periodontitis as well as increased levels of diabetes, SLE, and RA. The increased inflammation in each of these diseases provides a common framework for altering the oral microbiota [106].

In a recent longitudinal metagenomic analysis of the subgingival microbiome, diabetic patients were shown to be more susceptible to shifts in the subgingival microbiome toward dysbiosis, potentially due to impaired host metabolic and immune regulation. the subgingival microbiome in the healthy state in diabetic subjects had higher relative abundances of the orange complex and the red complex species than in non-diabetic subjects, suggesting an elevated risk of progression to periodontitis [107].

Higher bacterial load, a more diverse microbiota, an increase in bacterial species associated with periodontal disease, more clinical attachment loss, and increased production of inflammatory mediators including IL-17, IL-2, TNF, and IFN-γ were shown in RA patients. Furthermore, changes in the oral microbiota were linked to worse RA conditions [108].

Analysis of the oral cavity and its microbiome may be a useful tool to diagnose systemic diseases that have periodontal manifestations. Good oral hygiene to control the total microbial load is important to prevent dissemination to other body sites.

The Effect of Antibiotics Intake

The use of antibiotics is associated with an altered and often less diverse composition of the gut microbiome [109, 110]. However, potential negative effects of prophylactic antibiotic use on the oral microbiome have been entirely neglected. Salivary microbiome was found to be less affected and more resilient toward the exposure to antibiotics than the fecal microbial community [31]. Exposure to different antibiotics resulted change in microbial profiles that remained statistically significant compared to the placebo group for up to 4 months in the fecal samples while and the change was up to 1 month in saliva [31].

Unlike adults, early intake of antibiotics can have long-term consequences for microbiota development. In children, long-term alterations of the gut microbiome as a consequence of early antibiotic administration have been described and proposed to have negative effects for systemic health, including obesity and allergy [111, 112]. Significantly divergent colonizations were observed at 24 months and 7 years of age in children who did or did not take antibiotics during the first 2 years of life, whereas bacterial composition at earlier time points were overlapping in children treated with antibiotic in longitudinally collected oral samples in children followed from birth to 7 years of age [23]. *Actinomyces, Fusobacterium, Veillonella* and *Lactobacillus* were associated with antibiotics intake during the first 2 years of life, while *Neisseria* and *Streptococcus mitis/dentisani*, were present at significantly higher levels in 7-year-old children that did not take antibiotics [23]. The data suggest that the abundance of commensal genera such as *Granulicatella* [39] may be disturbed by antibiotics use, whereas the presence of other genera, like *Prevotella*, which has been associated with several oral diseases [58], may be favored.

Whether the use of broad-spectrum antibiotics might be a double-edged knife in the treatment of periodontitis is an issue worthy of discussion; such antibiotics will suppress also health-associated species and may thus enhance the dysbiotic status of the oral microbiome.

Characteristics of the subgingival microbiome at baseline and not the use of antibiotics was shown to predict the long-term (12 months) clinical outcomes of the non-surgical treatment of chronic periodontitis. Antibiotics resulted in a greater influence on the microbiome 3 months after therapy, but this difference disappeared at 6 months. Greater microbial diversity, specific taxa and certain microbial co-occurrences at baseline and not the use of antibiotics predicted better clinical treatment outcomes [113].

The Effect of Smoking on Oral Microbiota

Smokers show higher prevalence and severity of periodontal destruction and the therapy of periodontitis in smokers on average is less effective [114]. Smoking can affect the host immune response. Smokers with periodontitis show less serum antibodies, in particular immunoglobulin G class 2 (IgG2) and impaired function of leucocytes [115-117].

There have been conflicting reports on whether or not smoking has an effect on the periodontal microbiota. Some reported that smokers with periodontitis show higher prevalence and quantity of the traditional periodontitis-associated pathogens in comparison with non-smokers [118-120]; while others could not confirm those results [121, 122].

The microbial profile of smokers with periodontitis is distinct from that of non-smokers, using new techniques [123-125]. *Fusobacterium, Prevotella* and *Selenomonas* were more abundant in smokers, while *Peptococcus* and *Capnocytophaga* were more abundant in non-smokers. Low taxonomic diversity was associated with higher disease severity, especially in smokers [124].

CONCLUSION

The knowledge of microbiota in health and disease can provide an understanding of the pathogenesis of oral diseases, as well as changing and developing treatment strategies. Promoting a balanced microbiome is the key factor for maintaining and improving oral and general health.

REFERENCES

[1]   Marsh PD. 2018. "In Sickness and in Health-What Does the Oral Microbiome Mean to Us? An Ecological Perspective." *Adv Dent Res* 29(1): 60–65.

[2]   Filoche S, Wong L, Sissons CH. 2010. "Oral biofilms: Emerging concepts in microbial ecology." *J Dent Res* 89(1):8–18.

[3]   Marsh PD and Martin MV. 2009. *Oral Microbiology. Fifth ed.* London: Churchill Livingstone.

[4]   Diaz PI, Hoare A, Hong BY. 2016. "Subgingival microbiome shifts and community dynamics in periodontal diseases." *J Calif Dent Assoc* 44(7): 421–435.

[5]   Mira A, Simon-Soro A, Curtis MA. 2017. "Role of microbial communities in the pathogenesis of periodontal diseases and caries." *J Clin Periodontol* 44(18):23-38.

[6]   Lamont RJ, Koo H, Hajishengallis G. 2018. "The oral microbiota: dynamic communities and host interactions." *Nat Rev Microbiol* 16(12):745-759.

[7]   Mark Welch JL, Rossetti BJ, Rieken CW, Dewhirst FE, Borisy GG. 2016. "Biogeography of a human oral microbiome at the micron scale." *Proc Natl Acad Sci U S A* 113(6) : E791-800.

[8]   He X, Zhou X, Shi W. 2009. "Oral Microbiology: Past, Present and Future." *Int J Oral Sci* 1(2): 47–58.

[9]   Clarke JK. 1924. "On the bacterial factor in the etiology of dental caries." *Br J Exp Pathol* 5(3):141–147.

[10]   Fitzgerald RJ, Keyes PH. 1960. "Demonstration of the etiologic role of streptococci in experimental caries in the hamster." *J Am Dent Assoc* 61:9–19.

[11]   Zarco M F, Vess T J, Ginsburg G S. 2012. "The oral microbiome in health and disease and the potential impact on personalized dental medicine." *Oral Dis* 18(2): 109–120.

[12]   Kilian M, Chapple IL, Hannig M, et al. 2016. "The oral microbiome - an update for oral healthcare professionals." *Br Dent J* 221(10):657-666.

[13]   Ahn J, Yang L, Paster BJ, Ganly I, Morris L, Pei Z, Hayes RB. 2011. "Oral microbiome profiles: 16S rRNA pyrosequencing and microarray assay comparison." *PLoS One* 6(7):e22788. doi: 10.1371/journal.pone.0022788.

[14] Griffen AL, Beall CJ, Firestone ND. 2011. "CORE: A phylogenetically-curated 16S rDNA database of the core oral microbiome." *PLoS One* 22; 6(4):e1905. Published 2011 Apr 22. doi: 10.1371/journal.pone.0019051.

[15] Dewhirst FE, Chen T, Izard J, et al. 2010. "The human oral microbiome." *J Bacteriol* 192(19): 5002-5017.

[16] Escapa IF, Chen T, Huang Y, Gajare P, Dewhirst FE, Lemon KP. 2018. "New Insights into Human Nostril Microbiome from the Expanded Human Oral Microbiome Database (eHOMD): a Resource for the Microbiome of the Human Aerodigestive Tract." *mSystems*. 2018;3(6):e00187-18. Published 2018 Dec 4. doi: 10.1128/mSystems.00187-18.

[17] Chen T, Yu WH, Izard J, Baranova OV, Lakshmanan A, Dewhirst FE. 2010. "The Human Oral Microbiome Database: a web accessible resource for investigating oral microbe taxonomic and genomic information." *Database (Oxford)*. 2010;2010:baq013. Published 2010 Jul 6. doi: 10.1093/database/baq013.

[18] *Expended Oral Microbiome Database (eHOMD)*. 2020. Accessed May 06. http://www.homd.org/index.php.

[19] Dominguez-Bello MG, Costello EK, Contreras M, et al.2020. "Delivery mode shapes the acquisition and structure of the initial microbiota across multiple body habitats in newborns." *Proc Natl Acad Sci USA* 107(26), 11971–11975.

[20] Cephas KD, Kim J, Mathai RA, et al. 2011. "Comparative analysis of salivary bacterial microbiome diversity in edentulous infants and their mothers or primary care givers using pyrosequencing." *PLoS One* 6(8), e23503. doi: 10.1371/journal.pone.0023503.

[21] Jenkinsen HF, Lamont R. 2014. "Oral Microbial Ecology." In: *Oral Microbiology and Immunology, 2nd ed,* edited by Lamont RJ, Hajishengallis GN and Jenkinsen HF, 97-112. Washington DC: ASM Press.

[22] Gomez A, Nelson KE. 2017. "The oral microbiome of children: development, disease, and implications beyond oral health." *Microb Ecol* 73(2):492–503.

[23] Dzidic M, Collado MC, Abrahamsson T, et al. 2018. "Oral microbiome development during childhood: an ecological succession influenced by postnatal factors and associated with tooth decay." *ISME J* 12(9):2292-2306.

[24] Sulyanto RM, Thompson ZA, Beall CJ, Leys EJ, Griffen AL. 2019. "The Predominant Oral Microbiota Is Acquired Early in an Organized Pattern." *Sci Rep* 9(1):10550. Published 2019 Jul 22. doi: 10.1038/s41598-019-46923-0.

[25] Corby PM, Bretz WA, Hart TC, et al. 2007. "Heritability of oral microbial species in caries-active and caries-free twins." *Twin Res Hum Genet* 10(6):821–828.

[26] Kapil V, Haydar SM, Pearl V, Lundberg JO, Weitzberg E, Ahluwalia A. 2013. "Physiological role for nitrate-reducing oral bacteria in blood pressure control." *Free Radic Biol Med* 55:93–100.

[27] Kapil V, Weitzberg E, Lundberg JO, Ahluwalia A. 2014. "Clinical evidence demonstrating the utility of inorganic nitrate in cardiovascular health." *Nitric Oxide* 38:45–57.

[28] Jakubovics NS, Shields RC, Rajarajan N, Burgess JG. 2013. "Life after death: the critical role of extracellular DNA in microbial biofilms." *Lett Appl Microbiol* 57(6):467–475.

[29] Zhou Y, Gao H, Mihindukulasuriya KA, et al. 2013. "Biogeography of the ecosystems of the healthy human body." *Genome Biol* 14(1):R1. Published 2013 Jan 14. doi: 10.1186/gb-2013-14-1-r1.

[30] Rosier BT, Marsh PD, Mira A. 2018. "Resilience of the oral microbiota in health: mechanisms that prevent dysbiosis." *J Dent Res* 97(4):371-380.

[31] Zaura E, Brandt BW, Teixeira de Mattos MJ, et al. 2015. "Same exposure but two radically different responses to antibiotics: resilience of the salivary microbiome versus long-term microbial shifts in feces." *MBio* 6(6):e01693-15. Published 2015 Nov 10. doi: 10.1128/mBio.01693-15.

[32] Marsh PD, Moter A, Devine DA. 2011. "Dental plaque biofilms: communities, conflict and control." *Periodontol 2000* 55(1):16 –35.

[33] Zaura E, Keijser BJ, Huse SM, Crielaard W. 2009. "Defining the healthy 'core microbiome' of oral microbial communities." *BMC Microbiol* 9:259. Published 2009 Dec 15. doi: 10.1186/1471-2180-9-259.

[34] Bik EM, Long CD, Armitage GC, et al. 2010. "Bacterial diversity in the oral cavity of 10 healthy individuals." *ISME J* 4(8):962–974.

[35] Ling Z, Kong J, Jia P, et al. 2010. "Analysis of oral microbiota in children with dental caries by PCR-DGGE and barcoded pyrosequencing." *Microb Ecol* 60(3), 677–690.

[36] Palmer RJ Jr. 2014. "Composition and development of oral bacterial communities." *Periodontol 2000* 64(1):20-39.

[37] Chen H, Jiang W. 2014. "Application of high-throughput sequencing in understanding human oral microbiome related with health and disease." *Front Microbiol* 5:508. Published 2014 Oct 13. doi: 10. 3389/fmicb.2014.00508.

[38] Huse SM, Ye Y, Zhou Y, Fodor AA. 2012. "A core human microbiome as viewed through 16S rRNA sequence clusters." *PLoS One* 7(6):e34242. doi: 10.1371/journal.pone.0034242

[39] Aas JA, Paster BJ, Stokes LN, Olsen I, Dewhirst FE. 2005. "Defining the normal bacterial flora of the oral cavity." *J Clin Microbiol* 43(11):5721-5732.

[40] Seed PC. 2014. "The human mycobiome." *Cold Spring Harb Perspect Med* 5(5):a019810. Published 2014 Nov 10. doi: 10.1101/ cshperspect.a019810.

[41] Janus MM, Crielaard W, Volgenant CM, van der Veen MH, Brandt BW, Krom BP. 2017. "*Candida albicans* alters the bacterial microbiome of early *in vitro* oral biofilms." *J Oral Microbiol* 9(1):1270613. Published 2017 Jan 23. doi: 10.1080/20002297.2016. 1270613.

[42] Lepp PW, Brinig MM, Ouverney CC, Palm K, Armitage GC, Relman DA. 2004. "Methanogenic Archaea and human periodontal disease." *Proc Natl Acad Sci U S A* 101(16):6176-6181.

[43] Horz HP. 2015. "Archaeal Lineages within the Human Microbiome: Absent, Rare or Elusive?" *Life (Basel)* 5 (2):1333-1345.

[44] Marsh P D, Head D A, Devine D A. 2015. "Ecological approaches to oral biofilms: control without killing." *Caries Res* 49 (Suppl 1): 46–54.

[45] Richards VP, Alvarez AJ, Luce AR, et al. 2017. "Microbiome of site-specific dental plaque of children with different caries status." *Infect Immun* 85(8):e00106-17. Published 2017 Jul 19. doi: 10.1128/IAI. 00106-17.

[46] Jiang W, Ling Z, Lin X, et al. 2014. "Pyrosequencing analysis of oral microbiota shifting in various caries states in childhood." *Microb Ecol* 67:962–969.

[47] Topcuoglu N, Erdem AP, Karacan I, Kulekci G. 2019. "Oral microbial dysbiosis in patients with Kostmann syndrome." *J Med Microbiol* 68(4):609-615.

[48] Zaura E, ten Cate J M. 2015. "Towards understanding oral health." *Caries Res* 49 Suppl 1: 55–61.

[49] Marsh PD, Head DA, Devine DA. 2014. "Prospects of oral disease control in the future – an opinion." *J Oral Microbiol* 6:26176. Published 2014 Nov 27. doi: 10.3402/jom.v6.26176.

[50] Takeshita T, Kageyama S, Furuta M, et al. 2016. "Bacterial diversity in saliva and oral health-related conditions: the Hisayama Study." *Sci Rep* 6:22164. Published 2016 Feb 24. doi: 10.1038/srep22164.

[51] Mason MR, Preshaw PM, Nagaraja HN, Dabdoub SM, Rahman A, Kumar PS. 2015. "The subgingival microbiome of clinically healthy current and never smokers." *ISME J* 9(1):268–272.

[52] Wu J, Peters BA, Dominianni C. et al. 2016. "Cigarette smoking and the oral microbiome in a large study of American adults." *ISME J* 10(10): 2435-2446.

[53] Marsh PD. 1994. "Microbial ecology of dental plaque and its significance in health and disease." *Adv Dent Res* 8(2):263–271.

[54] Marsh PD. 2003. "Are dental diseases examples of ecological catastrophes?" *Microbiology* 149(Pt 2):279-294.

[55] Loesche WJ. 1986. "Role of Streptococcus mutans in human dental decay." *Microbiol Rev* 50(4):353–380.

[56] Caufield PW, Schön CN, Saraithong P, Li Y, Argimon S. 2015. "Oral lactobacilli and dental caries: a model for niche adaptation in humans." *J Dent Res* 94(9 Suppl):110–118.

[57] Bowden GH. 1997. "Does assessment of microbial composition of plaque/saliva allow for diagnosis of disease activity of individuals?" *Community Dent Oral Epidemiol* 25(1): 76–81.

[58] Aas JA, Griffen AL, Dardis SR, et al. 2008. "Bacteria of dental caries in primary and permanent teeth in children and young adults." *J Clin Microbiol* 46(4):1407-1417.

[59] Granath L, Cleaton–Jones P, Fatti LP, Grossman ES. 1993. "Prevalence of dental caries in 4- to 5-year-old children partly explained by presence of salivary mutans streptococci." *J Clin Microbiol* 31(1):66-70.

[60] Becker MR, Paster BJ, Leys EJ, et al. 2002. "Molecular analysis of bacterial species associated with childhood caries." *J Clin Microbiol* 40(3): 1001–1009.

[61] Takahashi N, Nyvad B. 2008. "Caries ecology revisited: microbial dynamics and the caries process." *Caries Res* 42(6):409–418.

[62] Tanner AC, Kressirer CA, Faller LL. 2016. "Understanding caries from the oral microbiome perspective." *J Calif Dent Assoc* 44(7):437–446.

[63] Henne K, Rheinberg A, Melzer-Krick B, Conrads G. 2015. "Aciduric microbial taxa including Scardovia wiggsiae and Bifidobacterium spp. in caries and caries free subjects." *Anaerobe* 35(Pt A):60–65.

[64] Mantzourani M, Gilbert SC, Sulong HN, et al. 2009. "The isolation of bifidobacteria from occlusal carious lesions in children and adults." *Caries Res* 43(4):308–313.

[65] Hajishengallis E, Parsaei Y, Klein MI, Koo H. 2017. "Advances in the microbial etiology and pathogenesis of early childhood caries." *Mol Oral Microbiol* 32(1):24-34.

[66] Eriksson L, Lif Holgerson P, Esberg A, Johansson I. 2018. "Microbial complexes and caries in 17-year-olds with and without Streptococcus mutans. " *J Dent Res* 97(3), 275–282.

[67]   Marsh PD, Zaura E. 2017. "Dental biofilm: ecological interactions in health and disease." *J Clin Periodontol* 44 Suppl 18: S12–S22.

[68]   Rosier BT, De Jager M, Zaura E, Krom BP. 2014. "Historical and contemporary hypotheses on the development of oral diseases: are we there yet?" *Front Cell Infect Microbiol* 4:92. Published 2014 Jul 16. doi: 10.3389/fcimb.2014.00092.

[69]   Kianoush N, Nguyen KA, Browne GV, Simonian M, Hunter N. 2014. "pH gradient and distribution of streptococci, lactobacilli, prevotellae and fusobacteria in carious dentine." *Clin Oral Investig* 18(2):659-69.

[70]   Kuribayashi M, Kitasako Y, Matin K, Sadr A, Shida K, Tagami J. 2012. "Intraoral pH measurement of carious lesions with qPCR of cariogenic bacteria to differentiate caries activity." *J Dent* 40(3):222-228.

[71]   Kianoush N, Adler CJ, Nguyen KA, Browne GV, Simonian M, Hunter N. 2014. "Bacterial profile of dentine caries and the impact of pH on bacterial population diversity. *PLoS One* 9(3):e92940. Published 2014 Mar 27. doi: 10.1371/journal.pone.0092940.

[72]   Anderson AC, Rothballer M, Altenburger MJ, et al. 2018. "In-vivo shift of the microbiota in oral biofilm in response to frequent sucrose consumption." *Sci Rep* 8(1):14202. Published 2018 Sep 21. doi: 10.1038/s41598-018-32544-6.

[73]   Nyvad B, Crielaard W, Mira A, Takahashi N, Beighton D. 2013. "Dental caries from a molecular microbiological perspective." *Caries Res* 47(2):89-102.

[74]   Belda-Ferre P, Alcaraz LD, Cabrera-Rubio R, et al. 2012. "The oral metagenome in health and disease." *ISME J* 6(1):46-56.

[75]   Nascimento MM, Browngardt C, Xiaohui X, Klepac-Ceraj V, Paster BJ, Burne RA. 2014. "The effect of arginine on oral biofilm communities." *Mol Oral Microbiol* 29(1):45-54.

[76]   Liu YL, Nascimento M, Burne RA. 2012. "Progress toward understanding the contribution of alkali generation in dental biofilms to inhibition of dental caries." *Int J Oral Sci* 4(3):135-140.

[77] Foxman B, Srinivasan U, Wen A, et al. 2014. "Exploring the effect of dentition, dental decay and familiality on oral health using metabolomics." *Infect Genet Evol* 22:201-207.

[78] Edlund A, Yang Y, Yooseph S, et al. 2015. "Meta-omics uncover temporal regulation of pathways across oral microbiome genera during in vitro sugar metabolism." *ISME J* 9(12):2605-2619.

[79] Papapanou PN. 2014. "Periodontal Diseases: General Concepts." In: *Oral Microbiology and Immunology, 2nd ed,* edited by Lamont RJ, Hajishengallis GN and Jenkinsen HF, 251-271. Washington DC: ASM Press.

[80] Meyle J, Chapple I. 2015. "Molecular aspects of the pathogenesis of periodontitis." *Periodontol 2000* 69(1): 7–17.

[81] Darveau RP. 2010. "Periodontitis: a polymicrobial disruption of host homeostasis." *Nat Rev Microbiol* 8(7): 481–490.

[82] Takahashi N. 2005. "Microbial ecosystem in the oral cavity: metabolic diversity in an ecological niche and its relationship with oral diseases." *Int Congr Ser* 1284:103–112.

[83] Hajishengallis G, Darveau RP, Curtis MA. 2012. "The keystone-pathogen hypothesis." *Nat Rev Microbiol* 10(10):717–725.

[84] Hajishengallis G. 2014. "The inflammophilic character of the periodontitis-associated microbiota." *Mol Oral Microbiol* 29:248–257.

[85] Duran-Pinedo AE, Chen T, Teles R, et al. 2014. "Community-wide transcriptome of the oral microbiome in subjects with and without periodontitis." *ISME J* 8(8):1659-1672.

[86] Herrero ER, Fernandes S, Verspegcht T, et al. 2018. "Dysbiotic biofilms deregulate the periodontal inflammatory response." *J Dent Res* 97(5):547–555.

[87] Yost S, Duran-Pinedo AE, Krishnan K, Frias-Lopez J. 2017. "Potassium is a key signal in host-microbiome dysbiosis in periodontitis." *PLoS Pathog* 13(6):e1006457. Published 2017 Jun 20. doi: 10.1371/journal.ppat.1006457.

[88] Nassar M, Tabib Y, Capucha T, et al.2017. "GAS6 is a key homeostatic immunological regulator of host-commensal interactions

in the oral mucosa." *Proc Natl Acad Sci USA* 114(3):E337-E346. doi: 10.1073/pnas.1614926114

[89] Abusleme L, Dupuy AK, Dutzan N, et al. 2013. "The subgingival microbiome in health and periodontitis and its relationship with community biomass and inflammation." *ISME J* 7(5):1016-1025.

[90] Socransky SS, Haffajee AD,Cugini MA, Smith C, Kent RL Jr.. 1998. "Microbial complexes in subgingival plaque." *J Clin Periodontol.* 25(2):134-44.

[91] Topcuoglu N, Kulekci G. 2015. "16S rRNA based microarray analysis of ten periodontal bacteria in patients with different forms of periodontitis." *Anaerobe* 35(Pt A):35-40.

[92] Curtis MA, Zenobia C, Darveau RP. 2011. "The relationship of the oral microbiotia to periodontal health and disease." *Cell Host Microbe* 10(4):302-306.

[93] Griffen AL, Beall CJ, Campbell JH, et al. 2012. "Distinct and complex bacterial profiles in human periodontitis and health revealed by 16S pyrosequencing." *ISME J* 6(6):1176–1185.

[94] Kumar PS, Leys EJ, Bryk JM, Martinez FJ, Moeschberger ML, Griffen AL. 2006. "Changes in periodontal health status are associated with bacterial community shifts as assessed by quantitative 16S cloning and sequencing." *J Clin Microbiol* 44(10):3665–3673.

[95] Kolenbrander PE, Andersen RN, Moore LV. 1989. "Coaggregation of *Fusobacterium nucleatum, Selenomonas flueggei, Selenomonas infelix, Selenomonas noxia*, and *Selenomonas sputigena* with strains from 11 genera of oral bacteria." *Infect Immun* 57(10): 3194-3203.

[96] Bradshaw DJ, Marsh PD, Watson GK, Allison C. 1998. "Role of *Fusobacterium nucleatum* and coaggregation in anaerobe survival in planktonic and biofilm oral microbial communities during aeration." *Infect Immun* 66(10): 4729-4732.

[97] Diaz PI, Zilm PS, Rogers AH. 2002. "*Fusobacterium nucleatum* supports the growth of Porphyromonas gingivalis in oxygenated and carbon-dioxide-depleted environments." *Microbiology* 148 (Pt 2): 467-472.

[98] Albandar JM, Susin C, Hughes FJ. 2018. "Manifestations of systemic diseases and conditions that affect the periodontal attachment apparatus: Case definitions and diagnostic considerations." *J Periodontol* 89 Suppl 1:S183-S203.

[99] Vandenberghe P, Beel K. 2011. "Severe congenital neutropenia, a genetically heterogeneous disease group with an increased risk of AML/MDS." *Pediatr Rep.* 2011;3 Suppl 2(Suppl 2):e9. doi: 10.4081/pr.2011.s2.e9.

[100] Tirali RE, Yalçınkaya Erdemci Z, Çehreli SB. 2013. "Oral findings and clinical implications of patients with congenital neutropenia: a literatüre review." *Turk J Pediatr* 55(3):241–245.

[101] Zaura E, Brandt BW, Buijs MJ, et al. 2020. "Dysbiosis of the oral ecosystem in severe congenital neutropenia patients." [published online ahead of print, 2020 Feb 5]. *Proteomics Clin Appl.* 2020; e1900058. doi: 10.1002/prca.201900058.

[102] Vieira AR, Albandar JM. 2014. "Role of genetic factors in the pathogenesis of aggressive periodontitis." *Periodontol 2000* 65(1):92–106.

[103] Eick S, Puklo M, Adamowicz K, et al. 2014. "Lack of cathelicidin processing in Papillon-Lefèvre syndrome patients reveals essential role of LL-37 in periodontal homeostasis." *Orphanet J Rare Dis.* 2014;9:148. Published 2014 Sep 27. doi: 10.1186/s13023-014-0148-y.

[104] Roberts H, White P, Dias I, et al. 2016. "Characterization of neutrophil function in Papillon-Lefevre syndrome." *J Leukoc Biol* 100(2):433-444.

[105] Jepsen S, Caton JG, Albandar JM, et al. 2018. "Periodontal manifestations of systemic diseases and developmental and acquired conditions: Consensus report of workgroup 3 of the 2017 World Workshop on the Classification of Periodontal and Peri-Implant Diseases and Conditions." *J Periodontol* 89 Suppl 1:S237-S248.

[106] Graves DT, Corrêa JD, Silva TA. 2019. "The oral microbiota is modified by systemic diseases." *J Dent Res* 98(2):148-156.

[107] Shi B, Lux R, Klokkevold P, et al. 2020. "The subgingival microbiome associated with periodontitis in type 2 diabetes mellitus." *ISME J* 14(2):519-530.

[108] Corrêa JD, Fernandes GR, Calderaro DC, et al. 2019. "Oral microbial dysbiosis linked to worsened periodontal condition in rheumatoid arthritis patients. *Sci Rep.* 2019;9(1):8379. Published 2019 Jun 10. doi: 10.1038/s41598-019-44674-6

[109] Clemente J, Ursell L, Parfrey L, Knight R. 2012. "The impact of the gut microbiota on human health: an integrative view." *Cell* 148(6):1258 –1270.

[110] Modi SR, Collins JJ, Relman DA. 2014. "Antibiotics and the gut microbiota." *J Clin Invest* 124(10):4212–4218.

[111] Reynolds LA, Finlay BB. 2017. "Early life factors that affect allergy development." *Nat Rev Immunol* 17(8):518–28.

[112] Ajslev TA, Andersen CS, Gamborg M, Sørensen TIA, Jess T. 2011. "Childhood overweight after establishment of the gut microbiota: the role of delivery mode, pre-pregnancy weight and early administration of antibiotics." *Int J Obes* 35:522–529.

[113] Bizzarro S, Laine ML, Buijs MJ, et al. 2016. "Microbial profiles at baseline and not the use of antibiotics determine the clinical outcome of the treatment of chronic periodontitis." *Sci Rep.* 2016;6:20205. Published 2016 Feb 1. doi: 10.1038/srep20205.

[114] Labriola A, Needleman I, Moles, DR. 2005. "Systematic review of the effect of smok-ing on nonsurgical periodontal therapy." *Periodontol 2000* 37, 124–137.

[115] Koundouros E, Odell E, Coward P, Wilson R, Palmer RM. 1996. "Soluble adhesion molecules in serum of smokers and non-smokers, with and without periodontitis." *J Periodont Res* 31: 596 –599.

[116] Graswinckel JE, van der Velden U, van Winkelhoff AJ, Hoek FJ, Loos BG. 2004. "Plasma antibody levels in periodontitis patients and controls." *J Clin Periodontol* 31(7):562-568.

[117] Rezavandi K, Palmer RM, Odell EW, Scott D A, Wilson RF. 2002. "Expression of ICAM-1 and E-selectin in gingival tissues of smokers

and non-smokers with periodontitis." *J Oral Pathol Med* 31(1):59–64.

[118] Haffajee A D, Socransky SS. 2001. "Relationship of cigarette smoking to the subgingival microbiota." *J Clin Periodontol* 28(5):377–388.

[119] Van Winkelhoff, AJ, Bosch-Tijhof CJ, Winkel EG, van der Reijden WA. 2001. "Smoking affects the subgingival microflorain periodontitis." *J Periodontol* 72(5):666-671.

[120] Gomes SC, Piccinin FB, Oppermann RV, Susin C, Nonnenmacher CI, Mutters R, Marcantonio RA. 2006. "Periodontal statusin smokers and never-smokers: clinical findings and real-time polymerase chain reaction quantification of putative periodontal pathogens." *J Periodontol* 77(9):1483–1490.

[121] Bostrom L, Bergstrom J, Dahlen G,Linder LE. 2001. "Smoking and subgingival microflora in periodontal disease." *J Clin Periodontol* 2:212–219.

[122] Van der Velden U, Varoufaki A, Hutter, JW, Xu L, Timmerman MF, Van Winkelhoff AJ, Loos B G. 2003. "Effect of smoking and periodontal treatment on the subgingival microflora." *J Clin Periodontol* 30(7):603–610.

[123] Shchipkova AY,Nagaraja HN, Kumar PS. 2010. "Subgingival microbial profiles of smokers with periodontitis." *J Dent Res* 89(11):1247–1253.

[124] Bizzarro S, Loos BG, Laine ML, Crielaard W. Zaura, E. 2013. "Subgingival microbiome in smokers and non-smokers in periodontitis: an exploratory study using traditional targeted techniques and a next-generation sequencing." *J Clin Periodontol.* 40(5): 483–492.

[125] Moon JH, Lee JH, Lee JY. 2015. "Subgingival microbiome in smokers and non-smokers in Korean chronic periodontitis patients." *Mol Oral Microbiol* 30(3):227-241.

INDEX

C

D

I

J

K

L

M

N